The Malachi Matrix

Individuals Sparking The Fire Of Revival

BY

Richard Peterson

Published by Book Writing Pioneer

Cover design by Book Writing Pioneer

ISBN: [978-1-964629-05-6]

BOOK WRITING
P I O N E E R

Dedication

*Dedicated to **Yehovah,***

God the Father.

Bringing Glory to His Name!

Declaration of Belief

My wife and I dedicate our lives to following Yeshua (Jesus) and
to serving Yehovah, God the Father.

Bringing Glory to His Name!

Our Mission

#1. To help believers BREAK FREE from Corporate Churches and Authoritarian Religious Hierarchies.

#2. Founding of "Elijah Day of the LORD Ministries."

#3. Help facilitate the formation of a Bible Based Fellowship of Believers

"The Malachi Matrix"

Individuals Sparking The Fire Of Revival

"And he shall turn the hearts of the father's to the children, and the hearts of the children to the fathers, lest I come and smite the earth with utter destruction."

Malachi 4:6.

Contents

Photo of the Author

Introduction

Throughout the history of man's relationship with the Creator God, Yehovah, there has been a reoccurring cycle among the people of God falling into idolatry and turning their backs on true worship. Yehovah God has consistently used INDIVIDUALS. He has chosen to spark the fires of revival. While large corporate Churches have held themselves up as doing "The Work of God," they have often become hierarchical and authoritarian. These same Churches have developed traditions that stifle spiritual growth. Each large Church has created a college to perpetuate its teachings. They mold students into obedient clones to be used as Church administrators and ministers. The number of study papers and booklets produced by these churches are mind numbing. These corporate approved materials are used to brainwash the ecclesia and reinforce the Church's "Statement of Beliefs". Over time, Statements of Belief come to be viewed as "The truth and nothing but the truth." Once you have this mindset, there is no room for spiritual growth or further revelation.

Statements of belief often hold more weight than scripture and must be defended at all costs. If anyone dares to question the Church's stated beliefs, he or she is viewed as a troublemaker or even a heretic. Some churches will feign the notion that the doctrine in question will be reviewed and responded to. The truth is they rarely, if ever, come back with an answer. If the person who brought the theological question persists in talking to other church members about it, or God forbid, starts quoting scriptural evidence, they are censored and told to never speak to other members about it again. If they refuse to be silent, they are threatened with dis-fellowship. All too often, they are cast out of the Church and even shunned by former friends and family members.

This book "The Malachi Matrix," will provide an archetype and a model for Courageous INDIVIDUALS who are tired of just playing Church. If you refuse to be censored any longer, if you long to make a difference in this world and in the world to come, fall on your knees and humbly ask Yehovah, God the Father, to light a fire in your soul and turn your heart back to His law and His Commanded forms of worship. I, for one, don't want to come to the end of my life filled with regrets. Wishing that I had done more to live like Jesus, Yeshua, and wishing I had done more to bring glory to the Father's Name. The days of wandering in a spiritual wilderness are over. It's time to turn your heart back to Yehovah, God the Father.

It's time to Come Back Home!

It's time to Embrace "The Malachi Matrix"!

Overview of the Book of Malachi

Malachi O Malachi! From now until the day I die, I will preach the word of Malachi (R.D.P. July 2022).

This tiny book of Malachi contains only four chapters. It is often overlooked, yet its message of "RECONCILIATION" is of paramount importance in these last days.

If ever a message needs to be shouted from the housetops, it is this:

Return to your Father!

Come home to Yehovah, your God!

6. For I am the LORD "Yehovah" I change not, therefore ye sons of Jacob are not consumed (burned up).

7. Even from the days of your Fathers, ye have gone away from my ordinances and have not kept them. Return unto Me, and I will return unto you saith the LORD "Yehovah" of hosts....

Malachi 3:6&7

I have created a chart on the following pages that will help us get a clear overview of all that is contained in the book of Malachi.

Book of Malachi Chart

Characters (Key Figures)	History (Time Frame)
#1. LORD- YHVH "Yehovah" Strongs 3068. Malachi 1:6, 3:6, 1:11.	The book of Malachi was written 450-400 BC during the time of Ezra Nehemiah. The 2nd temple existed.
#2. John the Baptist. Malachi 3:1 (first sentence) Matthew 11:7-11.	The 1st wave of Jews return from Babylon with Zerubbabel and Joshua 537 BC. Temple is rebuilt.
#3-A. Jesus "Yeshua". Lord in small caps. Adon or Adonai. Definition: sir, ruler, commander, master. Lord = Title LORD = Name	History falls silent. Religious backsliding. Intermarriage with foreigners. Ezra 9:12
#3-B. Jesus "Yeshua" Messenger of the covenant Malachi 3:1 beginning with the... the Lord whom you seek... Messenger in the Hebrew "Malek Brown Drivers Briggs #4397 Hebrews 8-6 better covenant" Hebrews 9:14-15 "new covenant" Malachi 3:1 "the covenant"	2nd & 3rd wave of Jews return from Babylon with Ezra & Nehemiah. 458-445 BC, during the reign of King Artaxerxes.
#4 Moses V "The law giver"	Priests and the people entered into a curse and an oath to do seven things. Nehemiah 10:29-39

#5 Elijah Malachi 4:546 Meaning of Elijah's name El-i-ah El is Yah My God is Yehovah	

Rebuke			Prophecy/Curses
Priests	**People**		
Refuse to honor Yehovah's name Malachi 1:6 & 1:11	Married Foreign wives. Malachi 2:11		Failure to give glory to Yehovah's name. Malachi 2:1-3
Married Foreign wives Malachi 2:11 Ezra 9:2	Neglected paying tithes and offerings Malachi 3:8-9		Corrupted Yehovah's law. Malachi 2:7-9
Offered sick and lame animals and polluted bread on the altar. Neglected the widows and the fatherless, Sorcery, Adultery, Mixing pagan worship Malachi 3:5	Failed to honor Yehovah as Father Malachi 1:6 & 4:6		Said serving Yehovah was vain and exalted the wicked and the proud. Malachi 3:14 & 15 malachi 4:1 & 5
			Prophecy/Blessings
			Blessings for glorifying Yehovah's name.

	Malachi 3:16-18
	Malachi 4:2-3
	Blessings for praying tithes and offerings Malachi 3:8-12

To begin our in-depth discussion on the book of Malachi, we need to mention this. We know almost nothing about the man Malachi from the Bible. There is no mention of lineage or his birthplace.. I did find this interesting information Concerning "Who was Malachi?"

The source was Chubad. org. Learning & Values, Jewish History by Avrohom Bergstein.

The title is: Who Was Malachi The Prophet?

And I quote:

Some opinions in the Talmud maintain that Malachi was Mordechi, the hero of the book of Ruth. Mordechi, it is thought, was referred to as "Malachi" because of his position as Viceroy of Persia. This title was a designation similar to that of an angel, or "Malek," who is a subordinate "Messenger of God." Another view is that Malachi is a pseudonym for Ezra, the scribe. A third view is that Malachi was neither Mordechi nor Ezra but a third person entirely.

End of quote.

In our chart under Characters (Key Figures), we find

#1. LORD- YHVH. Other than the prophet Malachi, there are only five individual beings identified in the book of Malachi. The most often mentioned individual is the LORD, God the Father. He is specifically mentioned by name 46 times. Once again LORD or LORD of hosts 46 times.

The translators render the tetragrammaton or four-letter name of God YHVH as LORD in large caps. This name, YHVH or LORD is from Strongs 3068.

Part of speech: Proper name.

Phonetic spelling: Yeh-Ho-Vah.

The definition is short, concise, and easy to understand.

"The proper name of the God of Israel".

I must ask, brethren, why is this simple definition so hard for so many in the Churches of God to understand?

#2. The second individual on our chart who is alluded to in the book of Malachi is John the Baptist.

Malachi 3:1 Behold I will send my messenger and he shall prepare the way before Me...

Stop.

How do we know that this single sentence from Malachi 3:1 is John the Baptist? Jesus "Yeshua" specifically told us so!

Matthew 11:7-10.

7. And as they departed Jesus "Yeshua" began to say unto the multitudes concerning John, what went you out into the wilderness to see? A reed shaken by the wind?

8. But what went ye out for to see? A man clothed in soft raiment? Behold they that wear soft clothing are in kings' houses.

9. But what went you out for to see? A prophet? Yes, I say unto you, and more than a prophet

10. for this is he whom it is written, Behold I send my messenger before thy face, which shall prepare the way before thee.

So, then, it is obvious that the first part of Malachi 3:1 is speaking of John the Baptist. By the way, I am discussing 5 key figures in order of their appearance in the text, not in order of their importance.

#3a. The 3rd individual alluded to in the book of Malachi is Jesus "Yeshua" {Lord in small caps).

Malachi 3:1 beginning with...and the Lord whom you seek shall suddenly come to his temple, even the messenger of the covenant whom you delight in, behold he shall come saith the LORD "Yehovah" of hosts.

The first time, this word Lord is mentioned in Malachi 3:1, it is in small caps. The translators got it right in the King James version here. The first word, Lord, in this verse is Adon or Adonai.

I found this in a resource titled: Adonai Definition and Meaning Dictionary.com. Adonai is the plural of the Hebrew word Adon, which means Lord or master. It was first used as God's title before it was used as God's name. {Stop). I might add it was first used as God's title before it was INCORRECTLY used as God's name.

So then, in Malachi 3:1, the first word Lord in lower case is "master" (a title)"...the Lord whom you seek."

The second word LORD in large caps is YHVH "Yehovah" {a proper name). "Thus, saith the LORD "Yehovah" of hosts.

Remember this simple explanation:

Lord in small caps - Title - "master"

LORD in large caps - Proper name - "Yehovah"

Think about this, my brethren. How often in the New Testament is Yeshua, Jesus, referred to as master?

#3-b Messenger of the Covenant.

Malachi 3:1 beginning with the Lord whom you seek shall suddenly come to his temple even the messenger of the covenant whom you delight in, he shall come saith the LORD "Yehovah" of hosts.

The messenger of the covenant mentioned here is translated as "angel of the covenant" in the Darby translation of the Bible. The word messenger here is "Malek" in Hebrew.

Brown Drivers Briggs Old Testament Hebrew-English Lexicon says of Malek #4397: From the unused root meaning to dispatch as a messenger from God, that is an angel. Also, a prophet, Priest, or teacher.

Almost every Bible commentary recognizes this "messenger of the covenant" as Jesus "Yeshua" the Christ. Malek is the standard word for messenger, both human and divine. Malekim plural refers to a group of supernatural beings as angels. Though all good angels are angels of God, there is one angel (messenger) who is special and unique from all other angels. He is called "the angel of God."

Exodus 14:19

And the angel of God which went before the camp of Israel, removed and went behind them, and the pillar of cloud went before their face, and stood behind them. Yehovah also called him "My angel."

Exodus 23:23

For "mine angel" shall go before thee, and bring you into the Amorites, and the Hittites, and the Perizzites, and the Canaanites, and shall cut them off. If we back up a few verses, we will have little doubt as to who "my angel" is referring to.

Exodus 23:20-21.

20. Behold I send an angel before thee, to keep you in the way, and to bring you into the place I have prepared.

21. Beware of him, obey his voice, provoke him not, for he will not pardon your transgressions, for My name is in him.

What does it mean here when it says, "My name is in him"? Please turn with me to Psalm 68:4

Psalm 68:4&5

4. Sing unto God: sing praises to His name: extol Him that rideth upon the heavens, by His name JAH and rejoice in His holy habitation.

5. A Father to the fatherless and a judge of the widow, is God in His holy habitation.

The name JAH (Yah) here is a contraction and the poetic name of Yehovah, God the Father. Yah is from Strongs H3065. The definition is the same as 3068 Yehovah: the proper name of the God of Israel.

Jesus Christ's given birth name in Hebrew is "Joshua" or, more accurately, "Yehoshua," which means Yah is salvation or Yehovah is salvation.

Luke 1:31-32.

31. And behold thou shalt conceive in thy womb and bring forth a son and shall call his name Jesus "Joshua".

32. He shall be great and shall be called the Son of the highest: and the LORD God shall give him the throne of his father David.

This explains how God the Father's name "Yehovah is in this angel mentioned in Exodus 23:20-21.

Further proof of the truth that this is a special angel or messenger is found in Hebrews 1:5

Hebrews 1:5 New King James

For to which of the angels did He ever say: "You are My Son, today I have begotten You"? And again: I will be to Him a Father, and He shall be to Me a Son"?

The only being who was in the role of a messenger (angel) and a servant from past eternity who Yehovah God bequeathed His holy and unique name "YAH" is Jesus "YAHSHUA".

It is worth mentioning that one of the "titles" of God the Father, "EL," is embedded in the names of three mighty archangels. Michael, Gabriel, and Helel. Most of us are familiar with Michael and Gabriel, but few know that Satan's name before the fall was Helel, which means:

Enchanting Luminary Spirit. God's revealed and unique name, "Yehovah," was only bequeathed to one being, the messenger or angel of Yehovah, who was named Yeshua the Christ.

Jesus "Yah-Shua" the Christ was present there with the children of Israel when they left Egypt and throughout their journeys to the promised land.

1 Corinthians 10:2-4.

2. And all were baptized unto Moses in the cloud and in the sea.

3. and did eat the same spiritual meat.

4. and did drink the same spiritual drink: for they drank of that spiritual rock that did follow them: and that Rock was Christ.

Jesus "Yeshua" was the Rock who followed them. Jesus "Yeshua" was the angel of Yehovah that went with them.

Now, I would like to share three witnesses that Jesus "Yeshua" is "the messenger of the covenant" and "the mediator of the covenant."

Hebrews 9:14-15.

14. how much more shall the blood of Christ, who through the eternal spirit, offered himself without spot to God, purge your conscience from dead works, to serve the Living god.

15. and for this cause He is the mediator of the new covenant, that by means of death, for the redemption of the transgressions that were under the first covenant, they which are called might receive the promise of eternal inheritance.

Jesus "Yeshua" the Christ, the Mediator of the New Covenant!

The second witness from scripture is found in.

Hebrews 8:6.

But now he has obtained a more excellent ministry, by how much also he is the mediator of a better covenant which was established on better promises.

Jesus, "Yeshua", Is the mediator of a better covenant.

For our third witness, turn in scriptures to

Malachi 3:1

behold I will send my messenger, and he shall prepare the way before me, and the Lord whom you seek shall suddenly come to his temple, even the messenger of the covenant, whom you delight in, behold he shall come sayeth the LORD Yehovah of host.

Jesus, "Yeshua", Messenger of the covenant.

We have already discussed the meaning of the word messenger. It is from the Hebrew word "Malek". To dispatch as a messenger from God, that is, as an angel. Now let us examine this title, Mediator. Mediator is from Strongs 3316 from the Greek word mesites, Pronounced Mah- see-tace. Definition: A go-between, intermediary, arbiter, agent of something good.

So, then Jesus, "Yeshua", is

1. The Messenger of the Covenant.

2. The Mediator of the Covenant.

3. The Mediator of a Better Covenant.

The fourth individual on our chart mentioned in the book of Malachi is Moses. Malachi 4:4. Briefly, Moses is known and often referred to as the lawgiver. He ascended Mount Sinai to meet with Yehovah, God the Father, and to receive the 10-commandment law of God, with all its statutes and judgments. Earlier on this very same mountain in the burning bush incident, God the Father, the God of Abraham, Isaac, and Jacob, revealed to Moses, his forever and eternal name, which is "YEHOVAH."

Exodus 3:15

And God said furthermore unto Moses, thus shall you say unto the children of Israel, the LORD "YEHOVAH" God of your fathers, the God of Abraham, the God of Isaac, and the God of Jacob hath sent me unto you: this is My Name Forever and this is My Memorial unto all generations.

The 5th and final individual in the left column of our chart under key figures is Elijah. He is only mentioned once, but like Moses he is mentioned by name.

Malachi 4:5&6.

5. Behold I will send you Elijah the prophet before the coming of the great and dreadful day of the LORD "Yehovah".

6. And He shall turn the heart of the fathers to the children and the heart of the children to the fathers, lest I come and smite the earth with total destruction.

I will have much more to say about Moses and Elijah concerning their prophetic mission, as described in the books of Malachi and Revelation in another chapter of this book.

For now, I'd like to explain the meaning of Elijah's name. Hebrew names are rich with meaning.

Elijah's name; El-i-Jah

Definition: El-is-Yah. My God is Yehovah!

Elijah's God is Yehovah, God the Father.

So, there we have it, brethren, the five key figures mentioned in the book of Malachi

#1. LORD- "Yah-Ho-Vah".

#2. John the Baptist.

#3. Jesus, Yeshua", messenger of the covenant.

#4. Moses- the law giver.

#5. Elijah the prophet.

LORD "Yehovah", God the Father, is mentioned 46 times. It is indisputable that the most important being in the book of Malachi is Yehovah, God the Father.

Now, I move to the second portion of our chart, which we have titled History/Time Frame. Just so you don't think I'm making up history, I need to mention my sources. Here are my cited works:

❖ McClintock and Strong Biblical Cyclopedia.
❖ The Encyclopedia of Ecclesiastical and Biblical Literature.
❖ Brown, Drivers, Brigg Hebrew Lexicon.
❖ Talmud and Apocrypha R. Travers Herford.
❖ New Jewish Encyclopedia.
❖ Strong's Concordance.
❖ Exegesis of the Book of Malachi by Malcum M. Hutton.
❖ Rabbinical Essays by Jacob Lauterbach.
❖ History of the Jewish People by Charles Foster Kent.
❖ Josephus: History of the Jews.
❖ Chabad. org. Jewish history by Avrohom Bergstein.

❖ Paper by Mr. Will Berg: 2020. Examining Church Governance.

❖ Finally, and most importantly, "His Story". The story of the history of man from Genesis to Revelation, "The Holy Bible".

I also need to voice this disclaimer concerning using historical commentaries. Some have recently stated that we should not use any historical material other than the Bible to glean information. The reason being that historical commentaries are written by men who often have trinitarian or other false religious leanings that are contrary to the truth of God as revealed in the scripture. I have also heard it stated that history is written by the victors of wars and conquers of nations, so their views are prejudiced towards their own race or nation. These are legitimate concerns and indeed should be considered. However! We do have a biblical principle that plainly states that when two or three witnesses agree on the details of an event, it is legitimately accepted as true testimony. We ignore extra-biblical history at our own peril. When we ignore corroborating histories only to prove our own religious interpretations of scripture, it is a dangerous path. I have also heard that it stated that without a basic historical context, it is nearly impossible to have a clear understanding of many of the stories laid out in scripture. To this concept, I fully agree.

Back to our chart on page 6 under: Historical/ Time Frame. The book of Malachi was written approximately from 450 to 400 BC, during the time of Ezra and Nehemiah. The second temple had been built and existed at this time.

We will need to discuss a brief history of Judah and the Jews during this time. If you remember, Judah went into captivity in Babylon in 597 BC. Jerusalem and was conquered by King Nebuchadnezzar. The temple and all the prominent houses in the city were burned. Judah ceased to exist as a nation at this time.

Exactly in Yehovah's perfect timing and in keeping with Isaiah's prophecy of 70 years of captivity, we have Cyrus, the great Persian king who conquers Babylon. The Jews in Babylon were commanded by Cyrus to return to Jerusalem, rebuild the temple as a house for the God of heaven, and to reestablish the true worship of Yehovah, the God of Israel.

Ezra 1:2.

Thus, sayeth Cyrus king of Persia, the Lord Yehovah, God of heaven hath given me all the kingdoms of the earth and he has charged me to build Him a house at Jerusalem, which is in Judah.

Only 42,360 Jews along with 7307 servants, maids, and male and female singers returned in this first wave. Most of the Jews remained in Babylon. Remember brethren that Jerusalem laid in ruins at this time. The Jews had done very well in Babylon some held important offices in government as leaders and advisers to the King. They were very prosperous in Babylon. They owned businesses. Many of the Jews simply said: "I don't want to go!"

It is interesting to know that even during the time of Christ, there were still far more Jews in Babylon than there were in Palestine.

So, in 537 BC, exactly 70 years after going into captivity, Cyrus, king of the Persians who had conquered Babylon, issued a decree stating that Yehovah, God of heaven, had charged him to build the House of God in Jerusalem.

He commissioned Jerubbaal as governor and Joshua as high Priest to be the leaders. It took 18 years after their return to rebuild the temple under constant siege from their enemy's roundabout. The second temple was dedicated in 516 BC. For about 50 to 60 years after the temple was dedicated and Passover was once again celebrated, history went silent. There is almost no record of what was happening in Judea at this time; we don't even know when

Zerubbabel and Joshua died. We only know that they both did die during this time frame and as often is the case when a strong leader or righteous King died in Judah, the people backslide in their devotion and worship.

Ezra 9:1&2.

1. Now when these things were done the Princess came to me saying, the people of Israel, and the Priest and the Levites, have not separated themselves from the lands, doing according to their abominations even the Canaanites, the Perizzites, the Jebusites, the Amorites, the Moabites, the Egyptians, and the Amorites.

2. For they have taken of their daughters for themselves, and for their sons, so that the holy seed have mingled themselves with the people of those lands. Yea the hand of the Princess and the rulers have been chief in this trespass.

An interesting side note to remember is that the story of Queen Esther in Babylon took place during this 50-to-60-year period of relative silence in Judea.

Ezra chapter 7 begins with, after these things in the reign of Artaxerxes.....

Artaxerxes was the Persian king of Babylon during the time of Ezra and Nehemiah. He reigned for 41 years. Artaxerxes made a decree to send a second wave of Jews to Jerusalem with Ezra, the scribe to reestablish the law of God.

Artaxerxes was apparently motivated by the God of heaven to give great favor and treasure to Ezra to fund his journey and his mission. He commanded Ezra to appoint magistrates and judges in Judea and Jerusalem.

Ezra 7:23-28

23. Whatsoever is commanded by the God of heaven let it be diligently done for the House of the God of heaven, for why should there be wrath against the realm of the King and his sons?

24. Also we certify you, that touching any of the Priest and Levites, singers, porters, Nethinims, or ministers of this House of God that it shall not be lawful to impose toll, tribute, or custom upon them.

25. And thou Ezra, after the wisdom of thy God, that is in thy hand, set magistrates, and judges which may judge all the people that are beyond the river Euphrates all such as know the law of thy God, and teach you them that know them not.

26. And whosoever will not do the law of God and the law of the king, let judgment be executed speedily upon him, whether it be on to death, or to punishment, or to confiscation of goods, or to imprisonment.

27. Blessed be the Lord "Yehovah," God of our fathers, which has put such a thing as this in the King's heart, to beautify the House of the Lord "Yehovah" which is in Jerusalem.

28. And has extended mercy unto me before the King and his counselors, and before the mighty Princess. And I was strengthened as the hand of the Lord "Yehovah" my God was upon me, And I gathered together out of Israel chief men to go up with me.

When Ezra arrives at Jerusalem in the summer of 458 BC. He finds people who have backslid in their worship. He also finds that they have intermarried with the pagan nations around them, contrary to the law of Moses.

Ezra 9:1&2, which we read earlier.

It is evident to me that the book of Malachi was written just prior to the time of Ezra and Nehemiah. As a prophetic warning, it must have been widely distributed to the priests and people in Judea, Jerusalem, and Babylon. Esther was still the Queen Mother during the beginning of Artaxerxes' reign. Esther, being a devout Jew, no doubt shared this prophecy with the Persian King of Babylon. How else do we explain the King's favor towards Ezra and Nehemiah? When Ezra arrived in Jerusalem, he busied himself diligently teaching the law of Yehovah and he had some success, but he still faced strong opposition from those who did not want to divorce their foreign wives.

So, then it wasn't until a third wave of Jews returned later with Nehemiah, who King Artaxerxes sent back to Judah and Jerusalem, that there was sufficient movement back to the true worship of Yehovah.

Nehemiah arrived in Jerusalem in the spring of 445 BC, with the blessing and full authority of King Artaxerxes. This was when the civil authority under Nehemiah the governor and the religious authority under Ezra the Scribe was strong enough to firmly establish the law and worship of Yehovah as the law of the land.

Finally, the problems inherent in marrying foreign people were resolved. Under the strong leadership of Nehemiah and Ezra, a plan was implemented to bring the priests and the people into compliance with the laws of God and the true worship of Yehovah.

The Priest and the people entered into an oath and a curse to do seven things.

Nehemiah 10:29-39

29. They clave to their brethren, their nobles, and entered into an oath to walk in God's law, which was given to Moses the servant of God, and to observe and do all the commandments of the LORD "Yehovah" and His judgements and His statutes.

30. and that we would not give our daughters unto the people of the lands, nor take their daughters for our sons.

31. and if any people of the lands bring any victuals on the sabbath day to sell that we would not buy it of them on the sabbath day: and that we would leave {observe} the seventh year and the exaction {forgiveness} of debt.

32. also made ordinances for us to charge the third part of a shekel for the service of the house of our God.

33. for the showbread, and for the continual meat offering, and for the continual burnt offering of the new moons, for the set feasts, and for the holy things, and for the sin offerings to make an atonement for Israel, and for all the work of the house of God.

34. And we cast lots among the Priests and the Levites, and all the people, for the wood offering. To bring it into the house of God, after the houses of the fathers, at times appointed year by year, to burn upon the altar of the LORD "Yehovah" our God as it is written.

35. And to bring the first fruit of our ground, and the first fruit of all fruit of the trees, year by year, unto the house of the LORD "Yehovah".

36. Also the firstborn of our sons, and of our cattle, as it is written in the law, and the firstlings of our herd and of our flocks, to bring to the house of our God. Unto the priests that minister in the house of our God.

37. And that we should bring the first fruit of our dough, and our offerings, and the fruit of all manner of trees, of wine, of oil, unto the priests, to the chambers of the house of our God: and the tithes of the ground unto the Levites that the Levites might have the tithes in all the cities of our tillage.

38. And the Priests, the sons of Aaron shall be with the Levites when the Levites take the tithes: and the Levites shall bring the tithes unto the house of God, to the chambers, into the treasure house.

39. For the children of the Levites shall bring the offerings of corn, of the new wine, and the oil, unto the chambers, where are the vessels of the sanctuary, and the Priests that minister, and the porters, and the singers, will not forsake the house of God.

So then, there were seven main things they vowed to do. They put it in writing. The Princess, government officials, and Priests, and religious officials signed this Constitution on behalf of the people. This group of leaders who signed later came to be known as: "The Great Assembly."

Nehemiah 9:38.

And because of all this we make a sure covenant, and write it; and our Princess, and Levites, and Priests seal unto it.

Here are the seven main things they vowed to do, and they put it in writing.

#1. They would keep the laws, judgements, and commandments of Yehovah.

#2. They would not intermarry with foreigners.

#3. They would observe the weekly Sabbath and all the annual holy days of Yehovah.

#4. They would observe the sabbatical year for resting the land and forgiving of debts.

#5. They would pay 1/3 shekel for the upkeep of the temple annually.

#6. They would supply all the wood for the altar of the temple.

#7. They would provide all the tithes and offerings commanded in the law.

It has been said of this time, after the building of the second temple and the reforms of Ezra and Nehemiah, that this was a time of spiritual revival for the Jews and perhaps the moment of highest fidelity to the keeping of the law and pure temple worship. This wonderful age of revival and fidelity lasted about 100 years.

So, we can say that Malachi's prophecies and message of reconciliation to Yehovah, God the Father, were successful in his day and during the time of Ezra and Nehemiah, but what about

Malachi 4:4-6.

3. Remember ye the law of Moses my servant, which I commanded unto him in Horeb, for all Israel, with the statutes and judgments.

4. Behold I will send you Elijah the prophet before the coming of the great and dreadful day of the LORD "Yehovah."

5. And he shall turn the heart of the fathers to the children and the heart of the children to the fathers, lest I come and smite the earth with utter destruction.

I contend the book of Malachi is a message for our day and our age because Elijah did not appear in Malachi's day. Will one like Elijah and one like Moses appear in our day? Stay tuned!

Now, we will discuss what is found on our chart under the headings Rebuke/Priests and Rebuke/People.

So, what do I mean by rebuke? These are the specific sins that Malachi points out to the priests and the people.

#1. Under "Rebuke/Priests" we find the subtitle: Refuse to Honor Yehovah's Name.

Malachi 1:6.

A son honored his father, and his servant his master: if than I be a Father, where is my honor, and if I be a master where is my fear? Sayeth the LORD "Yehovah" of hosts unto you O Priest that despise my name. And you say wherein have we despised thy name.

Malachi 1:11.

For from the rising of the sun even unto the going down of the same. My name shall be great among the gentiles: and in every place incense shall be offered unto My name, and a pure offering for My name shall be great among the heathen, saith the LORD "Yehovah" of host.

WOW! Isn't it amazing that the premier sin that Malachi points out to the priests and ministers is that they refuse to give honor to Yehovah's name. We all know from personal experience in our day that most priests and ministers in the churches absolutely refuse to give any credence to the Biblical truth that God the Father "Yehovah" Has a proper name. Here are a couple of shameful examples. Recent quotes by two different Church of God ministers.

#1. The name of God, "Yehovah," is a trivial issue.

#2. The name of God, "Yehovah," has no value. Now I ask you, what spirit is behind these statements?

Going forward, I hope to show that Malachi proves beyond any shadow of a doubt that there are severe penalties for not honoring Yehovah's name and wonderful blessings for honoring and upholding Yehovah's name.

Next on our chart under Rebuke/Priests, we have this rebuke: Married Foreign Wives Ezra 9:1&2.

1. Now when these things were done, the Princess came to me saying, the people of Israel, and the Priests, and the Levites have not separated themselves from the people of the lands, doing according to their abominations, even the Canaanites, the Hittites, the Perizzites, the Jebusites the Ammonites, the Moabites, the Egyptians, and the Amorites.

2. For they have taken of their daughters for themselves, and for their sons, so that the holy seed have mingled themselves with the people of the lands; yea the hand of the Princess and Rulers hath been chief in this trespass.

Ezra 10:9-12.

9. Then all the men of Judah and Benjamin gathered themselves together unto Jerusalem within three days. It was the 9th month on the 20th day of the month: and all the people sat in the street of the House of God, trembling because of this matter and for the great rain.

10. And Ezra the Priests stood up and said unto them, you have transgressed and have taken strange wives, to increase the trespass of Israel.

11. Now therefore make a confession unto Thee LORD "Yehovah God Of your fathers to do His pleasure; and separate yourself from the people of the land and from the strange wives.

12. Then all the congregation answered and said with a loud voice, as thou hast said, so must we do.

Notice once again that the Priests and the Princess were chief in this matter of marrying foreign wives, and as often as the case, the people followed in this trespass.

Next, under Rebuke/Priests, we find in our chart Sacrificed lamb and sick animals and offer polluted bread on the altar.

Malachi 1:7-10.

7. You offer polluted bread upon Mine altar, and you say wherein and have we polluted Thee? In that you say the table of the LORD "Yehovah" is contemptible.

8. And if ye offered the blind for sacrifice, is it not evil? And if ye offered the lame and sick is it not evil? Offer it now to the governor, will he be pleased with thee or accept thy persons sayeth the LORD "Yehovah" of hosts.

9. And now I pray you, beseech God that He will be gracious unto us, this has been by your means: will he regard your persons? Sayeth the LORD "Yehovah" of host.

10. Who is there even among you that would shut the doors for not? Neither do you kindle fire on Mine altar for not. I have no pleasure in you, sayeth the LORD "Yehovah" of hosts, neither will I accept an offering at your hand.

Concerning this detestable practice of offering lame and sick animals for sacrifice, here is some information I found in "Exegesis of the Book Of Malachi" Written by Malcolm M Hutton.

(And I quote:)

And the second Oracle of Malachi 1:6, 2:9, & 2:1-2. The prophet strongly denounces the Priest for their failure to give proper moral and spiritual guidance. The people's woes are apparently being explained because of a flaw in leadership. Yahweh asked his Priests where is the honor due for him? Honor that even the earthly child would unquestionably give his father. The law states that only spotless and unblemished animals were to be sacrificed to Yahweh. Yet the Priest had the gall to offer sick and maimed animals offerings they would not dare present before the governor. A curse was to fall on them for doing this. Yahweh had founded the priesthood upon Levi, who was the ideal of what a priest of Yahweh was supposed to be. In return for knowledge and the meaning of life, along with inward peace divinely given which came by virtue of his sacred office. The true Priest must stand in awe before the LORD "Yehovah" and reverence his holy name.

(End of quote)

I find it interesting that this was written by a Baptist minister.

And next, concerning polluted bread which was offered on Yehovah's altar, it is likely that they were offering stale or moldy bread, but I would contend that polluted bread may very well be a metaphor for false teachings. The pure bread which is undefiled could well be the words of Yehovah, God the Father, and the words of Jesus "Yeshua," The bread of life.

The polluted bread could well be the deception of false prophets and the false teachings of the Pharisees.

Finally, under Rebuke/Priests, On our chart, we have the subtitle: Neglected the Widow and the Fatherless. Along with this, we find sorcery, adultery, and mixing pagan worship.

Malachi 3:5.

And I will come near to you in judgment; and I will be a swift witness against the sorcerer, and against the adulterers, and against the false swearers, and against those who oppress the hireling of his wages, the widow, and the fatherless and that turn aside the stranger from his right, and fear not me sayeth the LORD "Yehovah" of hosts.

We see in this one verse a litany of sins. Sorcery had to do with the Pagan practice of using a medium to conjure up the spirits of the

35

dead. It was expressly forbidden in the law of Moses. I'll also state clearly here that this demonic practice stems from the satanic lie of the immortal soul. The sin of adultery was common among the worshippers of Ba-el which involved drunken orgies and even child sacrifice. Many of these sins were a natural outgrowth of marrying foreign wives and the mixing of Pagan forms of worship with the worship of Yehovah, the God of Israel.

Defrauding working men and women is listed as a sin here. How sad that this unfair practice is common in our day. Several times in the Bible, we read that Yehovah, the God of Israel, and Yeshua, the Lamb of God had a special place in their hearts for the widow and the fatherless. The phrase turned aside the stranger from his right had to do with turning a blind eye to the homeless the foreigner and the poor who often suffer in our society.

Now let us move to the next page of our chart under the subtitle: Rebuke/People.

Malachi 2:6.

Judah hath dealt treacherously, and an abomination is committed in Israel and in Jerusalem: for Judah hath profaned the holiness of the LORD "Yehovah" which he loved, and hath married the daughter of a strange God.

Yehovah calls this act of marrying foreign wives who worship strange gods an abomination! As I mentioned earlier, just like today, people tend to follow their religious and political leaders, in this case, marrying foreign wives. I do need to mention here unless some may say that Yehovah is racist or prejudiced. It is clear from scripture that Yehovah, the Father God is no respecter of persons. A mixed multitude of many races left Egypt with Moses and the children of Israel. Moses, it is recorded, married an Ethiopian woman who was dark skinned. It is clear from scripture that any man or woman of any race or nationality is welcome into the covenant of faith as long as they are willing to forsake the vain worship of false gods and embrace the true worship of Yehovah, God the Father.

Isaiah 56:6 N.K.J.

Also, the foreigner who joins themselves to the LORD "Yehovah" to serve him, and to love the name of the LORD "Yehovah", to be his servants. Everyone who keeps from defiling the Sabbath and holds fast My covenant.

Deuteronomy 10:17-19.

17. For the LORD "Yehovah" God is God of gods; and LORD of Lords, a great God, a mighty and terrible, which regards not persons, nor taketh rewards (bribes).

18. He doeth execute the judgment of the fatherless, and widow, and loveth the stranger in giving him food and raiment (clothing).

19. Love ye therefore the stranger, for you were strangers in the land of Egypt.

Matthew 22:35.

For I was an hungered and you gave me meat; thirsty and you gave me drink; I was a stranger and you took me in.

Moving on to our next rebuke found in the book of Malachi we see under the subtitle Rebuke/People

Neglected Paying Tithes and Offerings.

Malachi 3:8-9.

8. Will a man rob God? You have robbed me. But you're saying we're in have we robbed

Thee? In tithes and offerings.

9. You are cursed with a curse: for you have robbed Me, even this whole nation.

Just like nowadays, paying tithes and offerings in many churches has fallen to all-time lows. In my hometown of Waterloo, IA, it seems like one or two more long-standing churches every year just give up and shut their doors. Ministers earn so little that they can't get Americans to pastor churches. They bring in converts from Africa who are willing to live in America even on poverty wages. Here in the United States, about 30% of all adults identify as "Nones" people who never attend a Church of any kind. "Nones" simply means when surveyed about religious affiliation, they answer, 'none.'

The irony of this lack of giving and devotion is that many of us have experienced firsthand the incredible blessings that come from being faithful in tithes and offerings.

Finally, on our chart under Rebuke/People, we have:

Failed To Honor Yehovah as Father.

Malachi 1:6.

A son honors his father and a servant his master, if I then be a father, where is Mine honor? And if I be a master where is My fear? Sayeth the LORD "Yehovah" of hosts.

Malachi 4:6.

And he shall turn the heart of the fathers to the children, and the heart of the children to the fathers, lest I come and smite the earth with utter destruction.

The model prayer Jesus "Yeshua" gave about how to pray starts out "Our Father who art in heaven". Jesus "Yeshua" begins by acknowledging Yehovah as Father. Next, Yeshua says: Hallowed be thy name. In Hebrew, it goes like this; "Yet Kadesh Shem Kah." May your name be sanctified! Isn't that amazing! We are to sanctify and set apart as holy the name of God the Father, whose name is Yehovah!

With the influx of foreign wives during Malachi's day, bringing with them their gods, do you think perhaps titles they use for their gods begin to be substituted for the revealed name of Israel's God? A disturbing irony in the Pagan worship of Ba-el is that Ba-el was an honored title, meaning master or Lord. So, people back then and people today who carelessly assign the title Lord as a name for the Father God of Israel are belittling Yehovah by mixing His name with the Ba-els. Do you think Yehovah, the true God, just might be a little upset about this? Now we move on to the final portion of our chart on the book of Malachi Under "Prophecy Curses." The first subtitle is:

Failure To Give Glory to Yehovah's Name.

Malachi 2:1-3.

1. And now, O ye Priest (ministers) this commandment is for you.

2. If you will not hear, and if you will not lay it to heart, to give glory unto my name, sayeth the LORD "Yehovah" of hosts, I will even send a curse upon you, and I will curse your blessings yay I have cursed them already because you have not laid it to heart.

3. Behold, I will corrupt your seed and spread dung upon your faces, even the dung of your solemn feasts; and one shall take you away with it.

This stinging indictment is just as applicable today. The specific reason Yehovah sends a curse upon the Priests, think ministers today, is because they refuse to listen. "You will not lay it to heart"! Lay what to heart?

To give glory unto My name says the LORD "Yehovah" of hosts.

Sounds exactly like many ministers in the churches today. Let us read Malachi 2:2 one more time. If you will not hear, and if you will not lay it to heart, to give glory to My name saith YEHOVAH of hosts, I will even send a curse upon you, and will curse your blessings, yay I have cursed them already because you do not lay it to heart.

Yehovah, God the Father, warns anyone who refuses to give glory to his name that he will curse them. It is a fearful thing to fall into the hands of the living God!

Next, under the title: "Prophesy/Curses," we have: Corrupted Yehovah's Law.

Malachi 2:7-9.

7. For the Priest's {minister's) lips should keep knowledge, and they should seek the law at his mouth: For he is the messenger of the LORD "Yehovah of hosts.

8. But you departed out of the way: you have caused many to stumble at the law, you have corrupted the covenant of Levi, sayeth the LORD "Yehovah" of hosts.

9. Therefore have I also made you contemptible and base before the people, according as you have not kept my way, but have been partial in the law.

You have caused many to stumble!

You have not kept My ways!

You have been partial in the law!

How many churches today have compromised basic notions of right and wrong? How many have a kind of smorgasbord philosophy concerning the 10 commandments? They pick and choose which ones they will keep. Or worse yet they preach that the law has been done away. How many make no difference between clean and unclean? Holy and unholy? The curse here is that they will be made contemptible and base.

Very soon, when God the Father and Yeshua the Son of the living God return, it will be made very clear to everyone who is contemptible and base.

Next, under the title: "Prophesy/Curses," we have this: Said Serving Yehovah Was Vain and Exalted the Wicked and the Proud.

Malachi 3:14-15.

14. You have said it is vain to serve God: and what profit is it that we have kept his ordinances, and that we have walked mournfully before the LORD "Yehovah" of hosts?

15. Now we call the proud happy yea, they that work wickedness are set up; yea, they the temp God or even delivered.

Do we in our society today call the proud happy? We now have Gay (happy) Pride Parades and celebrations. How many in the radical "Woke" community are now appointed as government officials and elected to public offices, hired as newscasters and sports commentators? How many who have compromised or abandoned biblical views on sex and marriage are now being hired as ministers? "They that tempt God are delivered." How many that espouse nonbiblical views on sexuality are now protected (delivered) as a special protected social class?

Malachi 4:1. For behold the day cometh that shall burn as an oven; and all that are proud, yea all that do wickedly shall be stubble, and the day that cometh shall burn them up, sayeth the LORD "Yehovah" of host, that it shall leave them neither root nor branch.

How's that for a curse? Yehovah gives this warning: He says if you refuse to listen, if you refuse to turn your heart back to Yehovah, God the Father, the final curse for the incorrigibly wicked is that they will be thrown into the lake of fire and burned up. This is the second death, final and forever!

Revelation 20:12-15.

12 And I saw the dead small and great stand before God and the books were opened and another book was opened which is the book of life and the dead were judged out of those things which are written in the books, according to their works.

13 And the sea gave up the dead which were in it and death and hell delivered up the dead which were in them and they were judged every man according to their works.

14 And death and hell were cast into the lake of fire this is the second death.

15 And whosoever was not found written in the book of life was cast into the lake of fire. The final item we have under: "Prophesy/Curses" on our chart is this: Failed to Pay Tithes and Offerings.

Malachi 3:8-9.

8 Will a man rob God? Yet you have robbed Me, but you say wherein have we robbed Thee? In tithes and offerings.

9 You are cursed with a curse, for you have robbed Me, even this whole nation.

The final category on our chart about the Book of Malachi is hands down my absolute favorite. It is titled simply: "Prophesy/Blessings."

Oh, the blessings that flow when we surrender our lives, repent of our sins, and accept the atoning sacrifice of Yeshua, the Lamb of God. Oh, the blessings that are ours when we are reconciled to Yehovah, God the Father!

Malachi 3:16-18.

16. Then they that feared the LORD "Yehovah" spoke often one to another, and the LORD "Yehovah" hearkened and heard it, and a book of remembrance was written before Him, for them that feared the LORD "Yehovah" and that thought on His name.

17. And they shall be Mine, sayeth the LORD "Yehovah" of hosts, in that day when I make up my jewels: and I will spare them as a man spareth his own Son who serveth Him. 18. Then shall you return and discern between the righteous and the wicked, between him who serveth God and him who serveth Him not.

Malachi 4:2-3.

2 But unto you who fear My name, shall the Sun of righteousness arise, with healing in his wings and you shall go forth and grow up as calves of the stall.

3 And you shall tread down the wicked, for they shall be ashes under the soles of your feet in the day that I shall do this sayeth the LORD "Yehovah" of hosts.

May these wonderful blessings be ours! May all these prophetic words be said of us!

#1. That we revered Yehovah.

#2. That we thought upon His name.

#3. That our names are written in the book of remembrance in heaven.

#4. That we humbly and diligently discern between the righteous and the wicked, between those who serve Yehovah and those who do not serve Him.

#5. That we are counted as His jewels and as the sons and daughters of God the Father.

#6. That we endured until that day in which the sun of righteousness shall arise with healing in His wings to take us home.

#7. Finally, let it be said of us that we turned our hearts back to Yehovah, God the Father, and that we turned the hearts of the fathers to the children and the hearts of the children to the fathers.

This concludes our overview of The Book of Malachi.

I hope to show going forward how this book:

"The Malachi Matrix" and the book of Malachi is a Matrix, a Womb if you will, from which we may birth deeper revelations about Yehovah, God the Father, Yeshua, the Son of the Living God, and the message of reconciliation.

As I mentioned earlier, the book of Malachi is tiny, it only contains four chapters, but to quote Shakespeare: He may be small, but He be FIERCE!

The book of Malachi is a Matrix! Malachi's Eternal Message of RECONCILIATION is a paradigm for our day.

The Message of Reconciliation

Malachi 3:6&7.

6. For I am the LORD "Yehovah," I change not; therefore, ye sons of Jacob are not consumed (burned up).

7. even from the days of your fathers, ye have gone away from My ordinances and have not kept them. Return unto me and I will return unto you, saith the LORD "Yehovah" of hosts. As I mentioned in the introduction to this book throughout the history of man's relationship with the creator God "Yehovah," there has been a reoccurring cycle among the people of God falling into idolatry and turning their backs on true worship. Yehovah God has consistently used INDIVIDUALS He has called to spark the flames of revival and to preach the message of Reconciliation.

The Holy Bible is filled with true life stories of men and women just like you and me. Common people, ordinary people, who don't necessarily stand out in a crowd. INDIVIDUALS who can be described as humble, some even as timid. The deciding factor that set these INDIVIDUALS apart is that Yehovah, God the Father, called them and empowered them with the dynamic power of His holy spirit, enabling them to perform miracles, do wonderful work, and bring glory to Yehovah's name. We know from scripture that Yehovah, God the Father, does the calling of individuals.

John 6:44 Jesus "Yeshua" speaking.

No one can come to me unless the Father who sent Me draws him, and I will raise him up on the last day.

So, then, it is INDIVIDUALS who make up the Church.

In Acts 14:27, we read the following:

And when they were come and had gathered the "church" together, they rehearsed all that God had done with them and how He had opened the door of faith to the gentiles.

We need to examine the word "church." Here is the English definition of "church" from the Merriam-Webster Dictionary.

#1. A building for public and especially Christian worship.

#2. The clergy or officialdom of a religious body.

#3. {often capitalized) A body or organization of religious believers.

#4. A public divine worship.

#5. The clerical profession.

Here is the Greek definition of "church" from a source titled: Your Dictionary.

#1. The word "church" is from the Greek word Ekklesia, ecclesia in English.

#2. The members who make up the one ecclesia are not communities but individual men.

#3. The Greek ecclesiast means one who takes part in deliberations of the assembly, a debater, or a speaker in an assembly.

Notice the sharp contrast between the modern English definition of "church" and the original Greek definition. When I googled the word "church," I found this: The word translated "church" in the English Bible is "ekklesia." This word is from the Greek word "Kaleo" (to call) with the prefix "ek" [out]. Therefore, the word means "the called-out ones." The English word church does not come from ecclesia but from the word "Kuriakon," which means "dedicated" to the LORD. I also found this while googling the word "church." The oldest word for church, the word St. Paul himself used, is the Greek word 'ecclesia" from which we get "ecclesial" and "ecclesiastic." The word was in use centuries before the Christian church appeared on the scene.

As I stated earlier, there is a sharp contrast between the two meanings of the word "church." The English meaning, which is a corruption that occurred over thousands of years, emphasizes the following things:

#1. A building where believers meet.

#2. The clergy or officialdom of a religious body.

#3. Individual denominations such as Catholic, Methodist, or Baptist.

#4. The clerical profession.

#5. Finally, only timidly mentions a body of believers.

The Greek definition for 'church" ecclesia emphasizes the INDIVIDUAL! The members who make up the one ecclesia are not communities but INDIVIDUAL men and women. The

Greek ecclesiastics means one who takes part in the deliberations of an assembly. Ecclesia, a debater, or a speaker in an assembly.

The Greek word "church".

#1. Individual man or woman.

#2. One who takes part in deliberations.

#3. A debater.

#4. A speaker in an assembly.

#5. The called-out ones.

Let this sink into your heart and mind. When Jesus "Yeshua" spoke about His church, which of these two options was he thinking of? The obvious answer is "ecclesia." God the Father's church, which is founded on "The Rock" who is Christ, is made up of INDIVIDUAL man and woman who are actively involved. Taking part in deliberations, debating the meaning of scripture, and having a voice in the assembly. The called-out ones!

Yehovah, God the Father, and Jesus "Yeshua" work with each man and woman of the ecclesia one at a time. Salvation is not a group activity as in the Methodists shall inherit eternal life, or the Catholics shall inherit eternal life, or the Sabbath-keeping Churches of God shall inherit eternal life. INDIVIDUALS shall inherit eternal life! Salvation is extremely personal, and it is dependent on the INDIVIDUAL'S response to their calling and the choices they make. Only Yehovah God can save you through the redeeming blood of His Son, Jesus "Yeshua"—the Lamb of God who takes away the sin of the world.

John 1:29.

The next day, John seeth Jesus "Yeshua" coming unto him and saith, Behold the Lamb of God, who takes away the sin of the world.

How does Christ do that? One person at a time! The eternal relationship between Yehovah, God the Father, and Jesus "Yeshua" the Son of the living God, is personal. The relationship between Yehovah, God the Father, and each of His sons and daughters is personal. The relationship between each INDIVIDUAL believer and Jesus "Yeshua", the Messiah is personal!

Jesus "Yeshua" spoke to multitudes on several occasions, but it was INDIVIDUAL believers who responded to His message. We have numerous examples in the Bible of Jesus "Yeshua" speaking with INDIVIDUALS, one-on-one. The woman at the well is one such example. The woman at the well was a Samaritan. The traditions of the day demanded that Jews have no dealings with the Samaritans. Jesus "Yeshua" was not one to countenance manmade traditions. Yeshua had earlier sent his disciples away to a city to buy provisions. When the Samaritan woman approached, Yeshua asked her for a drink of water. What followed can only be described as a very comical verbal exchange between the snarky woman and Yeshua. The Samaritan woman said sarcastically, 'Are thou greater than our father Jacob which gave us this well, and drank from it himself? After Yeshua told her all the things she had done in her life, she had to admit that he must be a prophet. She then said this to Jesus: Our fathers worshipped on this mountain, but you say that Jerusalem is the place where men ought to worship. Jesus Yeshua answered her in John 4:21-24.

21. Jesus Yeshua sayeth unto her, "Woman, believe me, the hour cometh when ye shall neither on this mountain nor in Jerusalem worship the Father."

22. Ye worship ye know not what; we know what we worship for salvation is of the Jews.

23. But the hour cometh, and now is, when the true worshippers shall worship the Father in spirit and in truth, for the father seeketh such to worship Him.

24. God is spirit, and they that worship him must worship Him in spirit and in truth. Notice, brothers and sisters, Yeshua made it perfectly clear that the main object of true worship is Yehovah, God the Father!

He also made it clear that true worship is not dependent on a physical location or being a member of any particular group.

The temple in Jerusalem cannot save you. The Vatican in Rome cannot save you. Your church headquarters, wherever that may be, cannot save you. Salvation was then and is now wrapped up in being reconciled to Yehovah, God the Father. Jesus "Yeshua" was showing her that she needed to come back home. She needed reconciliation with Yehovah, God the Father.

Jesus "Yeshua" is worthy of worship and praise, but He always pointed to His heavenly Father as the focus of our worship. True worshippers shall worship the Father in spirit and in truth. Note this well! Your salvation is not dependent on your location on this earth or your membership in any corporate church.

WARNING!

The established churches of this world will not like the message contained in this book or me personally. They want to maintain their corporate authority and their hierarchal control over the ecclesia. They love to have the preeminence.3" John verses 9&10. They are addicted to the financial perks that come with their self-appointed offices. I will be accused of being a troublemaker and a heretic. I will be seen as causing division and not being a team player. The cry will go out that we must maintain unity. If you accept what I am teaching in this book and decide to free yourself from their domination over your spiritual growth, you will be accused of the same things.

Always remember this: your salvation as an INDIVIDUAL member of the ecclesia is personal. It has to do with being reconciled to Yehovah, God the Father, And worshiping Him in spirit and in truth.

Salvation is not a group activity or a team sport. I'm reminded of what the comedian and social critic George Carlin once said, and I quote: Somewhere along the way, someone is going to tell you, "There is no "I" in a team." What you should tell them is, maybe not, but there is an "I" in Independence, Individuality, and Integrity.

The whole book of Malachi is a matrix that shows us how we can be reconciled to Yehovah, God the Father. It's important that we address why I chose "The Malachi Matrix" as the title of this book. When we examine both these words, Malachi and Matrix, we will find a prophetic archetype going forward. The name Malachi means God's messenger. It can be human or divine, as in an angel. Building on this definition, it is easy to see how Yehovah, God the Father, uses INDIVIDUALS to carry his message of reconciliation to the world. The word matrix is a little more complex but rich with meaning.

So then, what does Malachi mean in English? Definition from Google:

Malachi is a masculine name of Hebrew origin, meaning messenger of God; it is a biblical name associated with the religious prophet and author of the book of Malachi. What does Malachi mean in Hebrew?

Definition from ABRARIM Publications- Biblical vault.

My Messenger, Messenger of Yah,

Angelic, angel of the LORD.

Now, the definition of the word Matrix is in Hebrew.

From Strongs concordance. Word #7358

Part of speech: noun masculine.

Pronunciation: rekh'hem".

Transliteration: rehem.

Definition: "The Womb".

What does the word Matrix mean in English?

From Merriam-Webster Dictionary

#1. Something within or from which something else originates. {the womb}?

#2. A mold from which a relief surface or type is made. {the womb}?

#3. Material from which something is enclosed or embedded. (the womb}?

Here are some examples of places where the word "matrix" is translated in the Bible as "womb."

Genesis20:18

For the LORD "Yehovah" had fast closed up the wombs (matrixes) of the house of Abimelech, because of Sarah, Abraham's wife.

Genesis 29:31

And when the LORD "Yehovah" saw that Leah was hated, He opened her womb (matrix), but Rachel was barren.

Exodus 13:2

Sanctify unto Me the firstborn, whatsoever openeth the womb (matrix) among the children of Israel, both man and beast; it is Mine.

The Hebrew definition of matrix is short, concise, and honestly easy to understand.

Matrix is from Strongs H7358, definition: "The Womb"! Is it any wonder that we have so many analogies in the Bible and directly from Jesus "Yeshua" about being "Born Again"? Here in our physical journey on this earth, the ecclesia (called out ones) enter the "Matrix" (The Womb"!)

The Apostle Paul made this eternal truth very clear in Galatians when he was comparing Sarah the freewoman to Hagar the bondwoman.

Galatians 4:25&26

25. For this Hagar is Mount Sinai, which is in Arabia, and answers to Jerusalem which now is, and is in bondage with her children.

26. But Jerusalem, which is above, is the mother of us all.

Now brethren, we have come to a very puzzling conundrum from scripture that must be resolved. Let us read the following scriptures from the book of Revelation, and while you are reading, ask yourself who or what is "The Bride of Christ?

Revelation 19:6-9

6. And I heard as it were the voice of a great multitude, and as the voice of mighty thunderings saying Hallelujah, for the LORD God omnipotent reigneth.

7. Let us be glad and rejoice and give honor to Him, for the marriage of the Lamb has come and His wife has made herself ready.

8. And to her was granted that she be arrayed in fine linen and white; for the fine linen is the righteousness of the saints.

9. And he saith unto me, Write Blessed are they which are called to the marriage supper of the Lamb. And he saith unto me these are the true sayings of God.

Now, carefully and thoughtfully read the following scriptures from

Revelation 21:9-11

9. And there came unto me one of the seven angels which had the seven vials full of the seven last plaques, and talked with me saying, Come hither and, I will show thee the bride, the Lamb's wife.

10. And he carried me away in the spirit to a great and high mountain, and showed me that great city, the holy Jerusalem, descending out of Heaven from God.

11. Having the glory of God and her light was like unto a stone most precious, even like a jasper stone, clear as crystal.

Now, go back and carefully read these two sets of scriptures once again. Slowly and thoughtfully.

Do you now see the conundrum? How can the faithful believers be the bride of Christ and, at the same time, New Jerusalem also be the bride of Christ?

The answer is they are both one and the same!

The new Jerusalem, which John saw in vision descending out of Heaven from God and revelation 21:10, is made-up of the ransom believers, the first fruits of the barley harvest who have been in Heaven with Christ during the tribulation.

Revelation 5:2-3

2. and I saw as it were a sea of glass mingled with fire: and them that had gotten the victory over the beast, and over the number of his name, stand on the sea of glass, having the harps of God.

3. And they sing the song of Moses, the servant of God, and the song of the lamb, saying, Great are they works, LORD God Almighty; Just and true are thy ways, thou King of saints.

When we combine this understanding with Galatians 4:26 But Jerusalem which is above is free, which is the mother of us all, it becomes apparent that the bride of Christ, the Lamb's wife, is destined to become the spiritual mother of an ever-expanding family of spirit born beings in the family of God the Father. Yehovah is the Supreme Godfather!

What then shall we say about Jesus "Yeshua"? Shall Christ also father children? How about this scripture?

Isaiah 53:10. Yet it pleased the LORD "Yehovah" to bruise him, to put him to grief: when Thou shalt make his soul an offering for sin, he (Christ) shall SEE HIS SEED, he shall prolong his days, and the pleasure of the LORD "Yehovah" shall prosper in his days.

In Isaiah 9:6, one of the titles given to Jesus "Yeshua is "the Everlasting Father." Do we now begin to see that Yehovah, God the Father, will always be referred to as the "Heavenly Father," just like Abraham is referred to as "Father Abraham," even by those who had Issac or Jacob or Judah as their actual fathers? Christ once said this:

John 5:19 Then answered Jesus "Yeshua" and said unto them, Verily I say unto you, the Son could do nothing of Himself. But what He seeth the Father do: for what things soever He doeth, these also doeth the Son likewise.

Yehovah, God the Father, is reproducing after His kind, spirit God beings!

Jesus, "Yeshua," is also reproducing after His kind, spirit God beings! The only reason for Christ to take a bride is for the purpose of PROCREATION. The bride of Christ will indeed become the New Jerusalem, the Mother of us all.

After the tribulation God the Father will return with Yeshua and all the saints to conquer the armies of this world. God the Father will make Christ's enemies His footstool.

Psalm 110:1 The LORD "Yehovah" said unto My Lord, "Adonai," Sit at My right hand until I make thy enemies Thy footstool.

Together, the family of God will establish the Kingdom of God on this earth. The Lamb of God "Yeshua" will present His Bride, New Jerusalem, to the entire world as the mother of us all. Revelation 21:9-11

9. And there came unto me one of the seven angels which had the seven vials full of the seven last plaques, and talked to me saying, Come hither and I will show you the bride, the lamb's wife.

10. And he carried me away in the spirit to a great and high mountain and showed me that great city, the holy Jerusalem, descending out of Heaven from God.

11. having the glory of God, and her light was like unto a stone most precious, even like a jasper stone, clear as crystal.

Yehovah, the Supreme Godfather.

Yeshua, the Everlasting Father.

The Bride of Christ, New Jerusalem, which will, from that day forward, be the Spiritual Mother of all future Children of God.

Jerusalem, which is above, could be none other than "New Jerusalem". The exact combination of the two words "New Jerusalem" only occurs twice in the Bible. Both times in the book of Revelation.

Revelation 3:12

Him that overcometh will I make a pillar in the temple of my God, and he should go no more out: and I will write upon him the name of my God which is new Jerusalem, which comes down from Heaven from my God, and I will write on him my new name.

Revelation 21:2

and I, John saw the holy city, new Jerusalem coming down from God out of Heaven, prepared as a bride adorned for her husband.

This bride spoken of here in revelation 21:2 are the believers who have been invited to the marriage supper of the lamb. The lamb is Yeshua of Christ who is the bridegroom.

There are times in the Bible when God refers to or speaks of things that shall be in the future as if they already are. Why? Because Yehovah is the sovereign God who will bring it to pass. When an individual person is called by Yehovah, God the Father, and accepts the blood of the Lamb of God, he is reconciled to God the Father and enters the matrix (The Womb), which is the fellowship of the ecclesia. We nurture one another with the word of God with serving one another, with comforting each other through our trials, with helping one another when we have physical needs. We are training to become the bride of Christ and a spiritual mother. The ecclesia looks above for its redemption. Our vision is to become the bride of Christ at the marriage supper of the lamb, which takes place in Heaven. Ultimately, if Christ takes a bride, what will the bride become? A MOTHER! The new Jerusalem is the future spiritual mother of a continuous, ever-expanding family of God. There is no reason for Christ to take a wife other than for the purpose of procreation.

Isaiah 66:7-13

7. Before she travailed, she brought forth, before her pain came, she delivered a man child.

8. Who hath heard such things? Who has seen such things? Shall the earth be made to bring forth in one day? Or shall a nation

be born at once? For as soon as Zion travailed, she brought forth her children.

9. Shall I bring to the birth and not cause to bring forth? Sayeth the LORD "Yehovah". Shall I cause to bring forth and shut up the womb (matrix) saith thy God.

10. Rejoice ye with Jerusalem and be glad for her, all you that love her, rejoice for joy with her all that mourn.

11. That you may suck and be satisfied with the breast of her consolation: that you may milk out and be delighted with the abundance of her glory. (See Revelation 21:11.)

12. For thus sayeth the LORD "Yehovah" Behold I will extend peace to her like a river, and the glory of the gentiles like a flowing stream; then shall you suck and be carried on her sides and be dandled upon her knees.

13. As one whom his mother comforts so will I comfort you, and you shall be comforted in Jerusalem.

May Yehovah give you the eyes to see this matter clearly. New Jerusalem! The Bride of Christ! Jerusalem which is from above! The Spiritual Mother of us all!

The Malachi Matrix

MALACHI: Yehovah's messenger. God's messenger.

You are invited to be reconciled to Yehovah, God the Father, and to become a messenger of God.

MATRIX: the bride of Christ becomes the new Jerusalem. Jerusalem from above. The mother of us all.

Enter the Matrix, safe within "The Womb" being nurtured and formed into a child of God. Awaiting the day, you are BORN AGAIN by a resurrection to ETERNAL LIFE!

Romans 8:19. For the earnest expectation of the creation eagerly waits for the revealing of the sons of God.

Isaiah 66:22. For just as the new heavens and the new earth which I make shall endure before me declares the LORD "Yehovah" so your offspring and your name shall endure. I ask you, my friends, as we look out at the vast universe which our creator God has stretched out to Infinity with the Hubble telescope and now the James Webb Space Telescope, we're able to gaze at distant galaxies and see into the past millions of years. As we marvel at the astounding beauty out there, don't you ever wonder what it is all for? Is it just as some have speculated, an endless meaningless void? Entertain this idea if you dare. Yehovah, the most high God, the LORD of hosts, the God and Father of our Lord Jesus Christ has this magnificent plan. He is PROCREATING! He is fathering a family of spirit God beings. Born again, sons and daughters. Could it be our mission to explore the universe? To seed life on distant worlds and planets?

Yehovah, God the Father, the creator God who spoke and brought the universe into existence. The one who breathed life into Adam and created Eve out of Adam's rib to be a wife, a helper, and a mother. He is the being who is worthy of all worship and all our praise. Jesus "Yeshua" is also worthy of our worship, but he is not above Yehovah, God the Father. God the Father is the Focal Point of Worship!

Christ is our example. Christ is the sacrifice, the Lamb of God, which his father offered. We emulate Christ and we follow Him as the Good Shepherd who is leading us back to God our Father. Christ is taking us home in the clearest sense of the word. As the bridegroom, He will come with all his holy angels to redeem His bride from a dying world. He will carry us to Heaven to the very throne of God Almighty. On the sea of glass, He will present us as the first fruits of the barley harvest to God the Father. Final preparations will be made to adorn the bride of Christ and for the marriage supper of the Lamb.

Revelation 21:2. And I John saw the holy city new Jerusalem coming down from God out of Heaven prepared as a bride adorned for her husband.

Jesus "Yeshua," at this very moment, is in Heaven preparing a home for His bride.

John 14:2-3.

2. In My Father's house are many mansions; if it were not so I would have told you, I go to prepare a place for you.

3. And if I go to prepare a place for you, I will come again and receive you to myself: where I am there you may be also.

Yeshua is the bridegroom!

John 3:28-29. John the Baptist speaking.

28. Ye yourself bear me witness that I said, I am not the Christ, but I am set before him.

29: He that hath the bride is the bridegroom, but the friend of the bridegroom, which standeth and heareth him, rejoiceth greatly because of the bridegroom's voice; this my joy is therefore fulfilled.

Just a personal conjecture on my part. At the marriage supper of the Lamb, who do you suppose might be Jesus's best man? The verse we just read says: but the friend of the bridegroom which standeth and heareth him... Jesus "Yeshua" once said, Truly I say unto you, among those born of women, there has not arisen anyone greater than John the Baptist. Could it be that Jesus "Yeshua" may choose John the Baptist to be His best man at the marriage supper of the lamb? Jesus "Yeshua" is, without a doubt, the bridegroom. Yehovah, God the Father officiates at the wedding, and God the Father invites INDIVIDUALS to the marriage of his Son.

Matthew 22:2-3.

2. The Kingdom of Heaven is like unto a certain king which made a marriage for his son.

3. And sent forth his servants to call them that were bidden to the wedding, and they would not come. The King, God the Father, sent forth His servants,

His Malachi, the messengers of God, to summon the invited guests.

Yehovah, God the Father, has a plan that originated in His mind. A plan that involves the procreation of sons and daughters. Yehovah Is fathering a family. The family of God!

2 Corinthians 6:18. And will be a father unto you, and you shall be my sons and daughters, sayeth the LORD Almighty.

Revelation 19:6,7 and 9

6. And I heard as it were the voice of a great multitude, and as it were the voice of many waters, and as the voice of many thunderings saying Hallelujah! for the LORD God omnipotent reigneth.

7. Let us be glad and rejoice and give honor to Him for the marriage of the Lamb has come, and His wife has made herself ready.

9. And he said unto me: Blessed are they which are called to the marriage supper of the Lamb. And he said unto me these are the true sayings of God.

So, then, do we only now begin to see the plan of God more clearly? Jesus "Yeshua" is the Lamb of God who was sent by Yehovah, God the Father. He suffered and died so that our sins could be forgiven. He is reconciling us to Yehovah, God the Father. Jesus "Yeshua" always pointed us to His Father as the one who has the highest honor.

John 12:44-45

44. Jesus "Yeshua" cried and said; He that believeth on Me, believeth on him who sent Me.

45. And he that seeth Me, seeth Him that sent me.

John 12:49-50.

49. for I have not spoken of Myself, but the Father who sent Me, He gave Me commandments what I should say, and what I should speak.

50. and I know his commandment is life everlasting, what so ever I speak therefore even as the Father said unto Me so I speak.

John 17:3 And this is life eternal that they might know Thee, the only true God and Jesus "Yeshua," whom You have sent.

Several times in my life, I have had well-meaning Christians ask me if I know the Lord. Meaning, do I know Jesus?

My question now to them is: Do you know the Father? Do you know His name? Jesus, "Yeshua," thought it was pretty important.

John 17:6 I have manifested THY NAME unto the men which Thou gavest Me out of the world; Thine they were and Thou gavest them Me; and they have kept thy word.

John 17:26 And I have declared unto them THY NAME and will declare it; that the love wherewith Thou hast loved Me may be in them and I in them.

Psalm 22:22. I will declare THY NAME unto my brethren: in the midst of the congregation will I praise Thee.

Proverbs 30:4 who has ascended into the heavens, or descended? Who has bound the waters in a garment? Who has established all the ends of the earth? What is His name, and what is His son's name if you know?

Psalm 83:18 That men may know that Thou, whose name alone is JEHOVAH are the most high over all the earth.

Once again, Jesus "Yeshua" said in John 17:3 "This is life eternal"! An integral part of receiving eternal life is knowing Yehovah, God the Father, and Jesus" Yeshua" the Christ who He has sent. The book of Malachi is all wrapped up in this same message. The message of RECONCILIATION to Yehovah, God the Father.

Malachi 3:6&7.

6. For I am the LORD "Yehovah" I change not; therefore, ye sons of Jacob are not consumed. (burned up)

7. Even from the days of your fathers, you have gone away from My ordinances and have not kept them. Return unto Me, and I will return unto you, saith the LORD "Yehovah" of hosts.

Yehovah, God the Father is calling you home. This dear friend is The Message of Reconciliation. The reason I chose this title "The Malachi Matrix".

MALACHI "Yehovah's Messenger"

God's Messenger.

You are invited to be reconciled to Yehovah, God the Father, and to become a messenger of God.

MATRIX "The Womb" The bride of Christ becomes "The New Jerusalem." Jerusalem from above. The Mother of us all.

Enter the Matrix. Safe within "The Womb," Being nurtured and formed into a child of God. Awaiting the day you are Born Again by a resurrection to eternal life.

The time for playing church is over! It is time to be reconciled to Yehovah, God the Father. It is time for the ecclesia to have a voice in the assembly of called out ones. If you want to make a difference in this life and in the world to come, join me in this ministry of reconciliation. Join me as we enter "The Malachi Matrix."

Wedding Supper of The Lamb/ In the Father's House

The book of Malachi is indeed a message about reconciliation. It is also a message teaching us to acknowledge, respect, and sanctify the eternal name of God the Father, which is Yehovah.

God the Father, Yehovah, is moving Heaven and earth to facilitate his plan of raising up the family of God. He has initiated a program that involves the procreation of eternal spirit God beings. Jesus, Yeshua, was sent by Yehovah as the messenger of the covenant. He introduced us to the revelation of an impending wedding. The parable of the marriage feast is instrumental in explaining several details concerning this wedding.

Matthew 22:1-14.

1. And Jesus answered and spoke unto them again by parables and said:

2. The Kingdom of Heaven is like unto a certain king, which made a marriage for his son.

3. And sent forth his servants to call them that were bidden to the wedding, and they would not come.

4. Again, he sent forth other servants, saying, tell them which are bidden behold I prepare my dinner my oxen and my fatlings are killed, and all things are ready, come unto the marriage. 5. But they made light of it, and went their ways, one to his farm, another to his merchandise.

6. And the remnant took his servants and entreated them spitefully and slew them.

7. But when the King heard thereof he was wroth; and he sent forth his armies, and destroyed those murderers, and burned up their city.

8. Then sayeth he to his servants, the wedding is ready, but they which were bidden were not worthy.

9. Go ye therefore into the highways, and as many as you shall find bid come to the marriage.

10. So those servants went out onto the highways and gathered together all as many as they found, both bad and good, and the wedding was furnished with guests.

11. And when the King came to see the guest, he saw there a man who had not a wedding garment

12. and he saith unto him, friend how comest thou in hither not having a wedding garment? And he was speechless.

13. Then said the King to the servants, bind him hand and foot and take him away and cast him into outer darkness there shall be weeping and gnashing of teeth.

14. For many are called, but few are chosen.

Several important details are revealed in this parable that speaks about the marriage of the King's son first, the King in the parable is obviously Yehovah, God the Father. He is the one who sends his servants to invite the guests to the wedding. His servants are all the righteous judges and prophets and ministers of God throughout history. God the Father, we understand is the one who calls people into the ecclesia of believers. Some have speculated that these are the Jews or others of the many tribes of Israel who have rejected the truth of God. I would contend that "those who are bidden" {invited) may include anyone who responded to the calling of Yehovah. Unfortunately, some later fall away from their calling notice in verse

5. But they made light of it and went their ways, one to his farm, another to his merchandise. There were also those who were called but were not totally committed. They were shirt-tail believers trusting and following a man. Many, even today, are trusting in the notion that they are the one true church, and they keep the Sabbath day, so they are good to go. Some have one foot in the world and one foot in the Kingdom. They are cavalier about their lifestyle and trust in their assumption of status with God. Then there are those who had the truth preached to them, which not only rejected Yehovah's servants, kicking them out of their synagogues (churches) and beating and killing some of them. When the time of the wedding arrived some of "those who were bidden" WHERE NOT WORTHY! I ask you, dear friends, how many people have you known personally who once were gung-ho about God's way of life who have turned around and walked away? The saddest words any man or woman could hear from Yehovah God the Father are found in Matthew 22:13. Then said the King, God the Father, to the servants, bind him hand and foot, and take him away, and cast him into outer darkness; there shall be weeping and gnashing of teeth. It is truly a fearful thing to fall into the hands of the living God.

The ministry of reconciliation is preached in the book of Malachi and is the essence of Jesus "Yeshua's" ministry. An often-overlooked detail of Christ's message is his revelations about the marriage supper of the lamb.

Revelation 19:1 and after these things I heard a great voice of much people, IN HEAVEN, saying Hallelujah; salvation and glory, and honor, and power unto the Lord our God. Revelation 19:6-7.

6. And I heard as it were the voice of a great multitude, and as the voice of many waters, and as the voice of mighty thunderings: saying Hallelujah for the Lord God omnipotent reigneth. 7. Let us be glad and rejoice, and give honor to Him, for the marriage of the LAMB has come, and HIS BRIDE has made herself ready.

As I pointed out in the previous chapter, Jesus "Yeshua" is the LAMB OF GOD. His BRIDE are those believers who He has raptured from the earth and taken to his Father's throne,

His Father's house. These faithful believers are presented to Yehovah, God the Father as the First Fruits.

Revelation 14:4. These are they which were not defiled with women, for they are virgins. These are they which follow the lamb wherever he goes, these were redeemed, (raptured) from among men, being the First Fruits unto God and to the Lamb.

These are they which were not defiled with women, for they are virgins.

This is a perfect lead-in to more revelation from Yeshua concerning the marriage supper of the lamb.

Matthew 25:1-13.

1. Then shall the Kingdom of Heaven be likened unto 10 virgins, which took their lamps and went forth to meet the bridegroom.

2. And five of them were wise and five were foolish.

3. They that were foolish, took their lamps, and took no oil with them.

4. But the wise took oil in their vessels with their lamps.

5. While the bridegroom tarried they all slumbered and slept.

6. And at midnight there was a cry made, Behold the Bridegroom Cometh, go ye out to meet him.

7. Then all those virgins arose, woke up, and trimmed their lamps.

8. And the foolish said unto the wise, give us of your oil, for our lamps are gone out. 9. But the wise answered saying not so; lest there be not enough for us and you, but go you rather to them that sell, and buy for yourself.

10. And when they went, the bridegroom came, and they that were ready went in with him to The Marriage and THE DOOR WAS CLOSED.

11. Afterwards came also the other virgins, saying Lord, open unto us.

12. But he answered and said truly I say unto you, I know you not.

13. Watch therefore, for ye know neither the day nor the hour wherein the son of man cometh.

The first thing we learned from this parable is that all 10 virgins initially expected the bridegroom to come for them and went out to meet him. Notice that the wise took oil with them. The foolish took lamps but took no oil with them. The oil, it would appear, represents the holy spirit, which is only provided by Yehovah and Yeshua. It is the spiritual force or power that enables the true servants of God to accomplish the works of preaching, teaching, healing, and serving the brethren. It is only given to those who truly repent and begin to live an obedient life. This oil is the spirit of God and the word of God that lights our path, enabling us to endure until the bridegroom cometh. The foolish virgins only put on an outward show of righteousness but continue to lead a double life. Only pretending to be holy at church while sinning on a regular basis in their private lives.

It is amazing to me that this scripture tells us that they all, all 10, slumbered and slept. It appears to this author they all initially were excited and expected the bridegroom to come by a specific date. It was only after the disappointment of what it refers to as "while the bridegroom tarried" that they all slumbered and slept. How many of us have experienced letdown when we trusted in those who told us Christ would come on such and such a date? There is, it would seem to me, very little of what we could call a spiritual revival or a great spiritual awakening happening today.

Next, in verse 6, it says, at midnight, there was a cry made, "BEHOLD THE BRIDEGROOM COMETH! Go ye out to meet Him! I must ask your brethren, when in the history of the modern church have you ever heard this message? BEHOLD THE BRIDEGROOM COMETH! In order to proclaim this message, one needs to acknowledge That Yehovah, God the Father, and Jesus, "Yeshua," the everlasting Father (Isaiah 9:6) are actively pursuing their plan of raising up the Family of God.

The next pivotal step in this plan is the marriage supper of the King's son. The wedding supper of the Lamb. The King is Yehovah, God the Father. The Son is Yeshua. Jesus, "Yeshua" is the Bridegroom. The hour we find mankind fast approaching is midnight.

There is a last-day Elijah message about reconciliation to Yehovah, God the Father, and the restoration of family, both physical and spiritual.

Jesus "Yeshua" is coming very soon to rescue (rapture) His Bride from a dying world. The shout needs to go out loud and clear: BEHOLD THE BRIDEGROOM COMETH!

For you to be counted as a wise virgin, you need to have sufficient oil in your vessel, and you need to ask for it.

The oil that Yehovah God provides is the spirit of God that will empower us to bring glory to His name. The key mistake the foolish virgins make is they do not acknowledge and look to Yehovah, God the Father, for the oil. They went to men; they trusted in men to save them.

The final lesson we need to learn from the parable of the 10 virgins in Matthew 25 is that time is running out! There is a day fast approaching when Christ will come for his bride and take her to Heaven, to the Father's house for the marriage supper.

Matthew 25:10-13 and while they went to buy, (the foolish virgins), the bridegroom came, and they that were ready went in with him to the marriage and THE DOOR WAS SHUT. The foolish virgins begged; Lord, open to us! But he answered and said, Truly, I NEVER KNEW YOU! Watch, therefore, for you know neither the day nor the hour were in the son of man cometh.

The marriage supper of the Lamb takes place in the Father's house. The Father's house is in Heaven. Jesus, Yeshua, said in

John 14:1-3

1. Let not your heart be troubled; do you believe in God, believe also in me.

2. In My Father's house are many mansions, if it were not so, I would have told you. I go to prepare a place for you.

3. And if I go and prepare a place for you, I will come again, and receive (rapture) you unto myself, so that where I am, there ye may be also.

Now I ask you, friends, where is Jesus right now? In Heaven! In the Father's House! Where will Yeshua the Bridegroom take His Bride for the marriage and the marriage supper of the Lamb?

Revelation 19:1 And after these things I heard a great voice of much people IN HEAVEN Say Hallelujah, Salvation and glory, and honor, and power, unto the Lord our God.

Revelation 19:5-9

5. And a voice came out of the throne saying, Praise our God, all ye His servants, and you that fear him, both small and great.

6. And I heard as it were the voice of a great MULTITUDE, and as the voice of many waters, and as the voice of mighty thundering's, saying Hallelujah; for the Lord God omnipotent reigneth.

7. Let us be glad and rejoice and give honor to Him for THE MARRIAGE OF THE LAMB HAS COME AND HIS BRIDE HAS MADE HERSELF READY.

8. And to her was granted that she should be arrayed in fine linen, clean and white; for the fine linen is the righteousness of the Saints.

9. And he sayeth unto me, Write Blessed are they which are called unto The MARRIAGE SUPPER OF THE LAMB. And he sayeth unto me, these are the true sayings of God. The true sayings of God make it abundantly clear that the marriage supper of the lamb takes place IN HEAVEN! IN THE FATHER'S HOUSE!

Those who are accounted worthy will be raptured by the Bridegroom, Jesus Christ, and taken to Heaven during the tribulation. We will be introduced to the King. Yehovah, God the Father. We will be wedded to Jesus "Yeshua" for all eternity. We will be instructed and taught how to be spirit beings. At the end of the tribulation, God the Father, Yehovah, along with Jesus "Yeshua," the Bridegroom and all the saints who are now the Bride of Christ, will return to earth to establish the Kingdom of God.

Revelation 19:11-15.

11 And I saw Heaven opened and Behold a white horse and He that sat upon him was called Faithfull and True, and in righteousness He doeth judge and make war.

12 And the armies WHICH WERE IN HEAVEN followed him, upon white horses clothed in fine linen white and clean.

15 And out of his mouth goeth a sharp sword, that with it He may smite the nations; and He shall rule them with a rod of iron; and He (Jesus "Yeshua") treads the winepress of ALMIGHTY GOD, (God the Father.)

Notice it says: the armies Which Were in Heaven follow Christ on white horses. This army of spirit beings is clothed in fine linen clean and white. Who is this Multitude that returns with Christ to fight the nations?

Revelation 19: 7-8

7. Let us be glad and rejoice and give honor to Him for THE MARRIAGE OF THE LAMB HAS COME and HIS BRIDE HAS MADE HERSELF READY!

8. And to her was granted that she should be arrayed in FINE LINEN CLEAN AND WHITE; for the fine linen is the righteousness of the saints.

The ARMIES that were in Heaven are all the saints which had previously been redeemed from among men. They were RAPTURED to Heaven by Christ.

There is no doubt that YEHOVAH, God the Father will come to earth along with Yeshua and all the saints.

Revelation 16:14 For they are the spirits of demons, working miracles, which go forth unto the kings of the whole world, to gather them together to THE BATTLE OF THE GREAT DAY OF GOD ALMIGHTY.

That Great Day of God Almighty is speaking of The Day of Yehovah, God the Father. Here is further proof that Yehovah, God the Father, is coming to dwell with his Family on the earth.

Revelation 21:1-5

1. And I saw a new heaven and a new earth, for the first Heaven and the first earth were passed away and there was no more sea.

2. And I John saw the holy city NEW JERUSALEM, coming down out of Heaven, prepared as a BRIDE ADORNED FOR HER HUSBAND.

3. And I heard a great voice out of Heaven saying: BEHOLD THE TABERNACLE OF GOD IS WITH MEN, AND HE SHALL DWELL WITH THEM, He shall be their God and they shall be His people and GOD HIMSELF shall be with them and BE THEIR GOD.

4. And God, God the Father, shall wipe away all tears from their eyes, and there shall be no more death, neither sorrow, nor crying, neither shall there be any more pain, for the former things are passed away.

5. And He that sat upon the throne said: BEHOLD I make all things new, And He said unto me: Write for these words are faithful and true.

YEHOVAH, God the Father, is coming down to this earth to live with His family. This has been His desire all along, to dwell with his children.

The title of this chapter is: Wedding Supper of The Lamb/In the Father's House. The Father's House is currently in Heaven. The day is coming when the Father's house will be on this earth. On that glorious day, Yeshua, the Bridegroom, will introduce His Bride to the whole world.

Revelation 21:9-11.

9 And then came unto me one of the seven angels having the seven vials full of the seven last plaques, and talked with me saying: Come hither, I will show you THE BRIDE, THE LAMB'S WIFE.

10 And he carried me away in the spirit to a great and high mountain, and showed me that GREAT CITY, NEW JERUSALEM, descending OUT OF HEAVEN from God.

11 Having the glory of God, and HER LIGHT was like unto a stone most precious, even like a jasper stone, clear as crystal.

There you have it, friends. The wedding supper of THE LAMB will be held IN HEAVEN! The BRIDE OF CHRIST is the redeemed (raptured) from among men. The faithful and true believers. There is no other reason for Christ to take a Bride other than for the purpose of PROCREATION! The BRIDE OF CHRIST becomes THE NEW JERUSALEM. The Mother of us all. The spiritual mother of all future born-again children of God who come forth from the Matrix.

We need to be about the business of overcoming and preparing to become The Bride of Christ. We need to study the word of God and pray for oil in our lamps to light our path until Yeshua, the Bridegroom comes to take us home to THE FATHER'S HOUSE!

Finally, the apostle Paul gave us this timeless admonition.

2 Corinthians 11:2

I am jealous for you with a Godly jealousy: for I have BETROTHED you to ONE HUSBAND so that Christ might present you as a pure virgin.

Christ will present us to God the Father, on the sea of glass, before the Father's throne, IN HEAVEN!

Oh, how I look forward to the day when we are presented to the whole world as the Bride of Christ, the New Jerusalem, the Spiritual Mother of the Family of God.

The Place of Safety/ In the Father's House

In this chapter, I will discuss the place of safety. As the title suggests, I believe the Holy Bible indicates that the place of safety during the tribulation will be in the Father's house. Much confusion has occurred because many in the several branches of the churches of God refuse to acknowledge there are altogether Three Separate Comings of the Messiah, who is Jesus "Yeshua" the Christ. In His FIRST COMING, Christ came as The Messenger of the Covenant and the Suffering Messiah in His prophetic role as the sacrificial Lamb of God. He is currently seated at the right hand of Yehovah, God the Father in Heaven.

Hebrews 10:12 But this man, after he had offered one sacrifice for sin forever, sat down on the right hand of God.

He will come back to this earth in his SECOND COMING as a bridegroom to rescue {rapture) his bride from a world that is about to plunge into the Great Tribulation. The true believers, who are the wise virgins, will be transformed into spirit beings and born again into the family of God. They will be presented to God the Father as the first fruits of mankind. They will be protected in heaven so that no harm comes upon them during the tribulation when satan and his demons are loosed on earth. They will have to be taught how to be spirit beings, and they will be married to Jesus Yeshua, who is the bridegroom for all eternity. They will become the New Jerusalem, which is the spiritual Mother of all future humans to be born again into the Family of God.

The THIRD and FINAL COMING of Jesus "Yeshua," the Messiah, will be when He comes with Yehovah, God the Father, as a conquering king to establish the Kingdom of God on earth. The

New Jerusalem will descend from God out of heaven and will be the world headquarters of the Wonderful World Tomorrow. The "olam-ha-ba" (The World to Come).

Now let us take a closer look at the THREE DISTINCT and SEPARATE comings of Christ in order.

#1. Yeshua's first coming is twofold.

A. Jesus "Yeshua" came the first time as the Malak or messenger of God.

B. Jesus "Yeshua" also came the first time as the Suffering Messiah and the sacrificial Lamb of God.

John 17:1-8

1. Jesus "Yeshua" spoke these words and lifted up His eyes to heaven, and said: Father, the hour has come. Glorify your Son, that your Son may glorify You.

2. As you have given him power over all flesh, that He should give eternal life to as many as You have given him.

3. And this is eternal life, that they might know You, THE ONLY TRUE GOD, and Jesus Christ who you have sent.

4. I have glorified you on the earth on the earth. I have finished the work which you have given Me to do.

5. And now, O Father, glorify Me with the glory which I had with You before the world was.

6. I have manifested Your name (YEHOVAH) to the men whom You have given Me out of the world. They were Yours, You gave them to Me, and they have kept Your words.

7. Now they have known that all things which You have given Me are from You.

8. For I have given them the words which You have given Me; and they have received them, and have known surely that it came from You; and they have believe that You sent me.

9. I pray for them. I do not pray for the world but for those whom You have given me; for they are Yours.

10. And all Mine (my seed) are Yours, and Yours are Mine, and I am glorified in them.

11. Now I am no longer in the world, and I come to You, Holy Father, keep through Your name (YEHOVAH) those whom You have given Me, that they may be one (family) as we are one.

12. While I was with them in the world, I kept them IN YOUR NAME (YEHOVAH). Those whom You have given Me I have kept, and none of them is lost except the Son of perdition that the scripture might be fulfilled.

John 12:49&50.

49. For I have not spoken on my own authority; but the Father (YEHOVAH} who sent me gave me a command, what I should say and what I should speak.

50. And I know His command is everlasting life. Therefore, whatever I speak, just as the Father has told me, so I speak.

Jesus Yeshua could not be clearer as to the fact that God the Father "Yehovah" sent Him into the world and gave Him every word He spoke. Jesus Yeshua is the Malik or messenger of God the Father Jesus "Yeshua" came the first time as a messenger of the covenant.

Malachi 3:1 Behold I will send my messenger, and he will prepare the way before me. And the Lord (master) whom you seek will suddenly come to His temple, even the messenger of the covenant and whom You delight. Behold He is coming, says the LORD "Yehovah" of hosts.

Jesus "Yeshua" came the first time as the Mediator of a Better Covenant.

Hebrews 8:6. But Now He, Yeshua, has obtained a more excellent ministry, inasmuch as

He is also a Mediator of a Better Covenant.

Jesus "Yeshua," in His first coming, came as Mediator of The New Covenant.

Hebrews 9:14&15

14 How much more shall the blood of Christ who through the eternal spirit offered Himself without spot to God, purge your conscience from dead works to serve the Living God.

15 and for this cause He is the Mediator of the New Covenant that by the means of death for the redemption of the transgressions that were under the first covenant, they which are called might receive the promise of eternal inheritance.

Yeshua Messenger of The Covenant

Yeshua Mediator of a Better Covenant

Yeshua Mediator of The New Covenant

B Jesus "Yeshua" also came the first time as the Suffering Messiah and The Sacrificial Lamb of God.

Isaiah 53:1-12

1. Who has believed our report? And to whom is the arm of the LORD "Yehovah" revealed?

2. For He (the Christ) shall grow up before Him (Yehovah) as a tender plant and as a root out of dry ground. He (Christ) has no form or comeliness; and when we see Him there is no beauty that we should desire Him.

3. He (Christ) is despised and rejected of men; a man of sorrows and acquainted with grief, we hid as it were our faces from Him: He was despised, and we esteemed Him not.

4. Surely, He has born our griefs and carried our sorrows, yet we did esteem Him stricken, smitten of God, and afflicted.

5. But He was wounded for our transgressions, He was bruised for our iniquities, the chastisement of our peace was upon Him; and with His stripes we are healed.

6. All we like sheep have gone astray; we have turned everyone to his own way; and the LORD "Yehovah" has laid on him the iniquities of us all.

7. He was oppressed and afflicted, yet He opened not His mouth: He is brought like a lamb to the slaughter, and as a sheep before his shearers is dumb, so He opened not his mouth. 8. He was taken from prison, and from judgement; and who shall declare His generation? For He was cut off out of the land of the living: for the transgression of the people was He stricken.

9. And He made his grave with the wicked, and with the rich in his death; because He had done no violence, neither was any deceit in his mouth.

10. Yet it pleased the LORD "Yehovah" to bruise Him, He has put him to grief: when Thou shall make his soul an offering for sin,

He shall SEE HIS SEED, He shall prolong his days, and the pleasure of the LORD "Yehovah" shall prosper in his hands.

11. He shall see the travail of His soul and shall be satisfied: by his knowledge shall My righteous servant justify the many: for He shall bear their iniquities.

12. Therefore I "Yehovah" shall divide the spoil with the strong; for He has poured out His soul unto death; and He was numbered with the transgressors, and He bare the sins of many, and made intercessions for the transgressors.

Once again, the whole chapter of Isaiah 53 makes it clear that Jesus" Yeshua" came the first time as The Suffering Messiah and as the humble Lamb of God who was sacrificed for the sins of all mankind.

#2. Jesus "Yeshua" shall come the SECOND TIME as THE BRIDEGROOM.

In His next mission to planet Earth, Jesus "Yeshua" will come as the Bridegroom to rescue (rapture) His bride from this world just prior to the tribulation. Yeshua will come back to this earth for His beloved and take her to heaven, to the Father's House.

There, His bride will be protected from satan and his demons that will be loosed on the earth. The Place of Safety, which is alluded to in several scriptures, is in The Father's House on the sea of glass before the very throne of Almighty God.

The cry will go forth: BEHOLD THE BRIDEGROOM COMETH!

Matthew 25:6

And at Midnight, there was a cry made: BEHOLD THE BRIDEGROOM COMETH! The important thing to remember about the second coming of Jesus, "Yeshua," is that no one knows the exact time this shall happen.

Matthew 24:36-44

36. But of that day or hour knoweth no man. No, not the angels of heaven, but My Father only.

38. For as in the days that were before the flood, they were eating and drinking, marrying and given in marriage until the day Noah entered into the ark.

39. And knew not until the flood came and took them all away, so also shall the coming of the Son of man be.

40. Then shall two be in the field; the one shall be taken and the other left.

41. Two women shall be grinding in the mill: one should be taken and the other left. 42. Watch; therefore, for you know not what hour your Lord Yeshua doth come.

I have no doubt that the redeemed from among men, those who are chosen to be the Bride of Christ, will hear an audible voice declare: BEHOLD THE BRIDEGROOM COMETH!

Notice several details that Jesus Yeshua reveals in this prophecy

#1. It will be a total surprise to most of mankind who are consumed with the pleasures and cares of this world.

#2. No one knows the exact day or hour of Christ's second coming as the bridegroom. No, not even the angels in heaven. No, not even Jesus Yeshua himself. Only Yehovah God the Father knows and decides when to send his Son to rapture His Bride.

#3. Some will be taken suddenly, and others will be left behind.

#4. The important thing to remember as the called and chosen of God, we need to maintain a continual state of readiness and godly lifestyle. We need to have oil in our lamps just as the wise virgins who are looking for their Bridegroom to come.

I need to point out that this idea of a secret rapture has been ridiculed and rejected by several churches of God. This author repents for having once accepted this teaching. The only way you can reconcile these scriptures such as one is taken and the other is left, and as in the days of Noah, is to acknowledge Jesus Yeshua comes the second time as the bridegroom to secretly rescue rapture his Bride, carrying her away to the Father's House in Heaven. The second coming is just prior to the tribulation. The redeemed saints will be taken to the place of safety during the tribulation. Revelation 3:10: because you have kept the words of my patience, I also will keep you from the hour of temptation (tribulation), which shall come upon all the world to try them that dwell on earth.

Philippians 3:20-21.

20. For our conversation (citizenship) IS IN HEAVEN, from which also we look for the savior, the Lord Jesus Yeshua the Christ.

21. who shall change our vile body, that it may be fashioned like unto His glorious body, according to the working, whereby He is able even to subdue all things unto Himself.

Zephaniah 2:2-3.

2. Before the decree brings forth, before the day pass as the chaff, before the day of the fierce anger of the LORD Yehovah come upon you, before the day of the LORD Yehovah's anger come upon you.

3. Seek the LORD Yehovah all ye meek of the earth, which have sought His judgement, seek righteousness, seek meekness, it

may be YE SHALL BE HID in the day of the LORD Yehovah's anger.

Psalm 91:1-2

1. He who dwells in the secret place of the Most High shall abide under the shadow of the Almighty.

2. I will say of the LORD Yehovah, He is my refuge, and my fortress, my GOD; in Him will I trust.

You cannot abide under the shadow of THE MOST HIGH GOD unless you are actually in His glorious presence. In the Father's House, in Heaven, on the sea of glass before his throne.

Revelation 19:5-7

5. and a voice came out of the throne saying: Praise our God, all you his servants, and all you who fear Him, both small and great.

6. And I heard the voice of a GREAT MULTITUDE, and as the voice of many waters, and as the voice of mighty thunderings saying: HALLELUYAH! For the God omnipotent reigneth.

7. Let us be glad and rejoice, and give honor to Him (Yehovah), for THE MARRIAGE OF THE LAMB has come, and HIS BRIDE has made herself ready.

Here we find MUCH PEOPLE and A GREAT MULTITUDE in heaven singing praises to God the Father before His throne. Shouting Hallelujah and Amen! How long before we admit and acknowledge the truth of what the Bible plainly tells us?

The place of safety is not in Petra, as some suggested in 1973 during the Yom Kippur War. The place of safety is not in Waco, TX. The place of safety is not with Jim Jones in the jungles of Guyana.

The place of safety for all the first fruits, the called-out ones, the wise virgins, THE BRIDE OF CHRIST, will be in heaven!

#3. The THIRD and final coming of Jesus Yeshua will be when He comes down to this earth with Yehovah, God the Father, to conquer the armies of this world and to establish the Kingdom of God on earth.

Revelation 1:7 Behold He is coming with the clouds, AND EVERY EYE SHALL SEE HIM, even those who pierced Him, and all the tribes of the earth will wail because of Him.

Matthew 24:30 Then shall appear in heaven the sign of the Son of Man, then shall all the nations of the earth mourn, AND THEY SHALL SEE THE SON OF MAN COMING IN THE CLOUDS WITH POWER AND GREAT GLORY!

Revelation 19:11-16

11. And I saw heaven opened and behold a White Horse: and He that sat upon him was called Faithful and true, and then righteousness does He judge and make war.

12 His eyes were as a flame of fire; on His head were many crowns, and He had a name written that no man knew, but He Himself.

13 and He was clothed were a vesture dipped in blood: and His name is called The Word of God.

14 And the armies (the redeemed) which WHERE IN HEAVEN followed him upon white horses, clothed in fine linen clean and white.

15 And out of His mouth goes a sharp sword, that with it He should smite the nations: and He (Jesus, Yeshua) shall rule them

with a rod of iron: and He treads the winepress of the fierceness and wrath of ALMIGHTY GOD!

16 AND He has on His vesture and on his thigh a name written, KING OF KINGS AND LORD OF LORDS!

Yehovah, God the Father, will have already poured out the seven vials filled with the seven last plagues upon the beast power and those who follow him. Notice what Yehovah, God the Father, prophesied in Psalms 110:1 The LORD Yehovah said unto my Lord (Adonai) sit thou at my right hand until I make thine enemies thy footstool.

Psalm 110:2-7

2 The LORD Yehovah shall send the rod of Thy strength out of ZION; rule Thou (Christ) in the midst of Thine enemies.

3 Thy people (the redeemed) shall be willing in the day of Thy power, in the beauty of holiness from the WOMB (Matrix) of the morning. Thou hast the dew of thy youth.

4 The LORD Yehovah has sworn and will not repent, Thou (Yeshua) art a Priest forever after the order of Melchizedek.

5 The Lord (Yeshua) at Thy right hand shall strike through kings in the day of His wrath. 6 He shall judge among the heathen, He shall fill the places with bodies, He shall wound the heads of many nations.

7 He shall drink of the brook (Gihon) in the way, therefore, He shall lop off their heads. Zachariah 14:4 is often quoted as referring to Jesus Yeshua, but the truth of scripture is that anywhere in Zechariah where the terms "The day of the LORD" or The LORD of hosts is used it is obviously referring to YEHOVAH, God the Father.

Zechariah 14:1 BEHOLD The day of the LORD Yehovah (God the Father) cometh and thy spoil will be divided in the midst of thee.

Revelation 16:14 For they are the spirits of demons, working miracles, which go forth unto the kings of the earth to gather them to the battle of THE GREAT DAY OF GOD ALMIGHTY!

Zechariah 14:3-4

3 Then shall the LORD Yehovah go forth and fight against those nations, as He fought in the day of battle.

4 And His feet (God the Father's feet) shall stand that day on the mount of olives which is in Jerusalem...

Jesus Yeshua is coming with his heavenly Father Yehovah and all the saints. He is coming the third time as a Warrior! The Lion of the tribe of Judah!

I am convinced that the world will be so deceived by the beast and the false prophet that they will believe this final coming of Yeshua, Yehovah, and the saints of God is an INVASION OF ALIENS! All this talk about UFOs and alien abductions is demonic in nature and convinces people that there are alien life forms out there in the universe who want to conquer our planet. Notice once again in Revelation 16:13-14.

13 I saw three unclean spirits (demons) like frogs come out of the mouth of the dragon (satan) and out of the mouth of the beast, and out of the mouth of the false prophet.

14 For they are THE SPIRITS OF DEVILS working miracles, which go forth unto the kings of the earth, and to the whole world, to gather them to the battle of THE GREAT DAY OF GOD ALMIGHTY.

The beast of revelation is a man who will soon arise in the not-too-distant future and who will be imbued with the very power and evil essence of Satan himself. The great false prophet, who is a religious leader of a great false church, will glorify this beast and convince the world to accept him as the savior of the world. Together, they will explain away the previous RAPTURE of the true believers (THE BRIDE OF CHRIST) as a MASS ALIEN ABDUCTION!

The two witnesses will preach a powerful message of Reconciliation to Yehovah, God the Father, during the last three and a half years of the great tribulation. The book of Revelation speaks of seven trumpet plaques along with seven last bowl plaques, which will be used to punish the beast and all those who follow him. The world will be convinced that it all has been the work of an ALIEN RACE. When Yehovah and Yeshua come with all the saints to establish the Kingdom of God on earth, the armies of the world will fight them to the death.

Know this, dear friend: the Bible speaks of a time when Satan and his demons came down to this earth to wreak havoc on mankind. This time is known as the time of Jacob's Trouble and the Great Tribulation. In order to be spared from this evil day, you need to humble yourself before Yehovah, God the Father. You need to repent of your sins, anything you have done which is contrary to the law of God and the Ten Commandments. You need to be reconciled to Yehovah, God the Father. You need to pay heed the invitation of Jesus, Yeshua, the Lamb of God.

John 14:6 Jesus Yeshua said unto him, I am the way, the truth, and the life; no man comes to the Father but by Me..

Matthew 11:27-30

27 all things are delivered unto Me of my Father, and no man knows the Son, but the Father, neither knows any man the Father save the Son, and he to whoever the Son will reveal Him.

28 Come unto Me all you that labor and are heavy laden (with sin), and I will give you rest.

29 take My yoke upon you and learn of Me, for I am meek and lowly of spirit and you shall find rest (forgiveness) unto your souls.

30 For My yoke is easy and My burden is lite.

Repent of your sins and ask forgiveness from Yehovah, God the Father. Accept the blood of Christ, the Lamb of God, to cleanse you from all your sins. Then, be baptized by full immersion as a symbol of the death of the old person. When you come up out of the water clean and pure, enter the MATRIX (THE WOMB), which is the body of believers. Read the Bible every day. Pray for the oil of God's holy spirit to sustain you, then go your way speaking about the love of God and strive to live a Godly life. Together, we will await that glorious day when Jesus, Yeshua, THE BRIDEGROOM, comes for His BRIDE. His Bride is YOU if you love the LORD, Yehovah, with all your heart, with all your mind, and with all your might. Keep your eyes fixed on Yeshua, the Good Shepherd, as He leads us to the Father's House. THE PLACE OF SAFETY!

The word Hallelujah is found only four times in the New Testament, all in Revelation chapter 19, verses 1-6. Hallelujah means Praise to Yah! Praise to Yehovah! Yehovah is God the Father. The word Amen is used 8 times in Revelation. Amen means: It is True or So be It. The words Hallelujah and Amen are pronounced the same in every language on earth.

I hope to see you in The Father's House, in THE PLACE OF SAFETY! We will join the chorus of angels singing the song of Moses and the song of the Lamb. We will proclaim glory and honor and power unto the LORD, Yehovah, and unto the Lamb.

HALLELUJAH AND AMEN!

The Focal Point of Worship

If someone were to ask you, as a Christian, who do you worship, what would be your answer?

Some would say: I worship Jesus Christ. Another may say: I worship God the Father. Still another might answer: I worship God the Father and Jesus Christ. What if we were to ask Jesus Christ the same question? Who did Jesus, Yeshua, worship?

John 20:17 Jesus Yeshua saith unto her, touch me not: for I have not ascended to My Father, but go to my brethren and say unto them, I ascend unto My Father and your Father, unto My God and your God.

Jesus Yeshua referred to Yehovah, God the Father, as My God and your God.

John 17:3 and this is life eternal, that they might know Thee, the only true God, and Jesus Christ whom Thou hast sent.

John 10:29 my Father which gave them to Me is greater than all, no man is able to pluck them out of My Father's hand.

Jesus said his Father is The Supreme God. The Most High God. "My Father is Greater Than All."

Did Paul and the apostles understand that Jesus, Yeshua, had a God?

1 Peter 1:3 Blessed be the God and Father of our Lord Jesus Christ, which according to His abundant mercy has begotten us again unto a lively hope by the resurrection of Jesus Christ from the dead.

Peter said: Blessed be the God and Father of our Lord Jesus Christ.

So, then brethren, if Jesus, Yeshua, called his Father: "the only true God." If Yeshua referred to His Father as "My God and your God, " If the apostles knew that God, the Father was Jesus' God. If Yeshua said: My father is greater than all" then " who is the Being we should be worshipping?

Clearly, Yehovah, God the Father, is the God we should worship.

Well, some will argue aren't Jesus, Yeshua, and God the Father one? The same God being? Some even try to teach that Jesus, Yeshua, is the God of the Old Testament. In the Bible, when we have a description of the throne, whether it's in heaven or on earth, there are two individual God beings pictured. Yehovah, God the Father, and Jesus, Yeshua, the Son of the Living God. God the Father, The Almighty God, is always center stage, and Jesus, Yeshua, is always at His right hand. Acts 7:54-56.

54. When they heard these things, they were cut to the heart, and they gnashed on him with their teeth.

55. But he, being full of the Holy Spirit, looked up steadfastly into heaven and saw the glory of God, and Jesus, Yeshua, standing at the right hand of God.

56. And said: behold I see the heavens opened and the Son of man standing at the right hand of God.

Stephen saw Two God Beings!

Revelation 21:22 and I saw no temple therein for the Lord God Almighty and the Lamb are the temple of it.

Clearly, the Lord God Almighty is God the Father, and the Lamb is Jesus, Yeshua. Once again, Two Eternal God Beings.

Revelation 22:3 And there shall be no more curse, but the throne of God and of the Lamb shall be in it: and his servants shall serve Him.

Two Thrones and Two Individual God Beings.

It is interesting that we never ever in scripture see a description of a third God being.

That's because the doctrine of the Trinity is complete fiction. Honestly, it comes from Pagan ideas about a triad god or god family.

Below, I give examples of Pagan triad gods or god families:

Greek Olympic triad gods: Zeus (king of gods), Athena (goddess of war), Apollo (god of the sun}

Roman Capitoline triad gods: Jupiter (the supreme god), Juno (his consort), Minerva (his daughter)

Ancient Egypt Triad: Osiris (husband) Isis [wife] Horus (son)

Celtic Mythology: Esas, Touatis Taranis.

Norse Mythology: Odin, Vili, Vi.

So, you see, brethren, the Pagan ideas of a triad, our trinity of god beings and god families, was very strong in ancient cultures, influencing false ideas about the nature of Israel's God. These Pagan ideas have corrupted the true worship of Yehovah, God the Father.

This we do know that being known as Satan was a fallen Angel, or Cherib, who desires to be worshiped and is trying to pass himself off as a God. Satan was puffed up and enamored with his own beauty. Satan managed to deceive and convince 1/3 of all the angels to join him in rebellion against the Most High God, Yehovah. Satan was cast out of heaven and his fall is symbolically described in Isaiah 14:12-14.

12. How art thou fallen from heaven O Lucifer, son of the morning (shining star). How art thou cut down to the ground, where thou weakens the nations.

13. For thou hast said in thine heart I will ascend into heaven; I will exalt my throne above the stars of God: I will sit also upon the mount of the congregation in the sides of the north.

14. I will ascend above the heights of the clouds: I will be like The Most High. Satan wants to be worshipped.

Matthew 4:10-11

10. And he, (satan), said to him, all these things I will give you if you will fall down and worship me.

11. Then Jesus, Yeshua, said to him, away with you Satan, for it is written, you shall worship the LORD, Yehovah, and Him only shall you serve.

And now, we come to the crux of the matter, The Focal Point of Worship. Yehovah is the revealed name of the God of Abraham, Isaac, and Jacob. He is the Being that spoke to Moses from the midst of the burning Bush.

Exodus 3:15. And God said moreover (furthermore) unto Moses, thus shall you say to the children of Israel. The LORD, YEHOVAH, God of your father's, the God of Abraham, the God of Isaac, and the God of Jacob, has sent me unto you. This is MY NAME FOREVER, and MY MEMORIAL unto all generations.

YEHOVAH is the proper name of the God of Israel! The Creator God! The one Jesus, Yeshua, called My God and your God; My Father and your Father. John 20:17.

Some want to argue that Jesus, Yeshua, is the being that spoke to Moses from the burning bush and from Mount Sinai. Jesus Yeshua told us plainly that God spoke to Moses and called him "He."

Mark 12:26&27

26. And as touching the dead that they rise: have you not read in the book of Moses how in the Bush God spoke to him, saying I am the God of Jacob?

27. He is not the God of the dead but the God of the living. You, therefore, do greatly err.

Jesus did not say: "I spoke to Moses". Jesus, Yeshua, was not confused about who the God of the Old Testament was. Jesus, Yeshua was indeed a God being that existed with Yehovah, God the Father in past eternity.

John 17:5 and now O Father, glorify Thou Me with Thy own self with the glory which I had with thee before the world was.

Jesus, Yeshua, didn't receive his name until Gabriel, an angelic messenger, revealed the name to Mary. Jesus, Yeshua, who was born as a baby in Bethlehem, never once in his life on earth heard anyone call him Jesus. His actual given birth name was Yehoshua. This is a Hebrew name that means Yehovah is salvation. Over time, Yahoshua was contracted to Yeshua.

In most Christian churches today, Jesus is the focus of worship. Jesus, Yeshua, is emphasized in songs. Pictures of Jesus are everywhere. Usually, the worshipper will pick the picture or painting of Jesus that appeals to their own emotions. Many today pick what can only be described as a handsome Jesus. At Easter time, some are drawn to the suffering Jesus with a grimace on his face. Some recently portrayed Jesus as a rugged Jew. Some choose the skin tone of their own race. Perhaps one of the most popular portrayals nowadays is the "glorified Jesus." He is often glowing and surrounded by an army of angels. The best prophecy about the coming Messiah found in Isaiah chapter 23 clearly describes Jesus, Yeshua, as very ordinary in appearance.

Isaiah 53:2 He shall grow up before him as a tender plant and as a root out of dry ground. He hath no form nor comeliness, and when we shall see Him, there is no beauty that we should desire him.

All of man's images of Jesus, Yeshua, are used to exalt and worship him. There are very few believers who ever consider whether all these portrayals are graven images.

One thing is undeniable, you have seldom, if ever, seen a picture or painting of God the Father. The only ones I have ever seen, and they are rare. God the Father looks like an ancient white-haired Olympian God like Zeus or Neptune. He looks very much like Charlton Heston's portrayal of Moses at the end of the classic 10 Commandments movie.

Replacement theology, also known as supersessionism, is a Christian theology that describes the belief or conviction that the Christian Church has superseded the Jews and the nation of Israel, assuming their role as God's covenanted people. Though most who hold this view won't admit it, they also believe that Jesus Christ is a loving, compassionate, and meek Messiah who has superseded Yehovah, God the Father, who is viewed as the harsh, wrathful God of the Old Testament. They also teach that Jesus, Yeshua, did away with the law.

Matthew 5:17. Think not that I came to destroy the law, or the prophets. I came not to destroy but to fulfill.

The word fulfill here is from the Greek word ple-roo. Strongs #4137 To cram, level up, execute, verify, and fully preach.

Matthew 5:18 Whosoever therefore shall break one of the least commandments, and shall teach men so, he shall be called least in the Kingdom of heaven, but whosoever shall do and teach them, the same shall be called great in the Kingdom of heaven.

Jesus, Yeshua, would teach that not only is the letter of the law still binding on believers, but the spirit of the law is binding as well. If we entertain an evil thought and dwell on it, we are as guilty as if we committed the act. It's the thought that precedes and precipitates the action. The spiritually evil thought needs to be brought into submission.

Finally, brethren, who is the one that Jesus, Yeshua, held up as worthy of our love, devotion, and worship? The first and greatest commandment is found in Exodus 20:2&3.

2. I am the LORD, Yehovah, thy God.

3. Thou shall have no other gods before me.....

In verse 2, God the Father introduces himself and tells us His name. In verse 3, He states: You shall have no other gods ABOVE Me, AHEAD of Me, OTHER than Me! Was Jesus, Yeshua, a God being before his human existence? Yes, He Was!

Is the risen Jesus Christ once again a glorified God being seated at the right hand of Yehovah, God the Father in heaven? Yes, He Is! So then, can Jesus, Yeshua, come before or above Yehovah, God the Father?

NO, HE CAN NOT!

Mark 12:28-30 &32.

28. And one of the scribes came and having heard them reasoning together and perceiving the answer well, asked him, which is the first (great) commandment of all?

29. And Jesus, Yeshua, answered him, the first (greatest) of all the commandments is, Hear O Israel; the LORD, Yehovah, our God is one LORD.

30. And thou shalt love the Lord thy God with all thy heart, and with all thy soul, and with all thy mind, and with all thy strength; this is the first (great) commandment.

32. And the scribe said unto him, Yeshua, Well Master has Thou said the truth; for there is one God and there is none other but He.

Jesus, Yeshua, as was his custom, quoted from scripture when he answered the scribe. He quoted from

Deuteronomy 6:4-5

4. Here O Israel; The LORD Yehovah, our God is one LORD.

5. And thou shalt love the LORD, Yehovah, with all thine heart, with all thy soul, and with all thy might.

So once again, brethren, who is the being that Jesus, Yeshua, said is the one God who is worthy of our complete love, devotion, and worship?

YEHOVAH! God the Father!

The Being revealed in the holy Bible is the FOCAL POINT of WORSHIP! He is the one whom Jesus, Yeshua, prayed to, saying: Hallowed be Thy name. In Hebrew it is: "Yet-Kah- Desh-Shem-Kah." May Your name be sanctified. By now, many who have read this far may be thinking, there he goes again, obsessing about "The Name."

I would like to lay down a challenge. Do a study of the book of Psalms and the book of Isaiah looking for every place the word "name" and the word "LORD" are used together in the same verse. Keep in mind that the word LORD is a terrible English translation of the Hebrew YHVH or Yehovah from Strongs #3068, the proper name of the God of Israel.

Isaiah 47:4 As for our Redeemer, the LORD, Yehovah of host, is His name the holy one of Israel.

Psalm 29:2. Give unto the LORD, Yehovah, the glory due His name; worship the LORD, Yehovah, in the beauty of holiness.

Psalm 34:3. O magnify the LORD, Yehovah, with me, and let us exalt His name together.

And now, my friends here are two verses from your Bible where even the King James translators got the name of God the Father right. Well almost.

The "J" is silent. The "J" is from a German derivation.

Isaiah 12:2 Behold God is my salvation: I will trust and not be afraid: for the LORD, Jehovah, is my strength and my song, He also is my salvation.

Psalm 83:18. That men may know that Thou whose name alone is Jehovah art the most high over all the earth.

One of the things I love about the book of Malachi is that it focuses on "YEHOVAH," God the Father, and it makes it clear that His Eternal Name is of paramount importance for the true worshipper. Those who will worship the Father in spirit and truth.

Yehovah, God the Father, longs to dwell with His children here on earth. He is coming down to this earth with Jesus, Yeshua.

Joel 3:16. The LORD, Yehovah, also shall roar out of Zion, and utter His voice from Jerusalem; and the heavens and earth will shake; but the LORD, Yehovah, will be the hope of His people, and the strength of the children of Israel.

Zachariah 14:9. And the LORD, Yehovah, shall be king over all the earth; in that day shall there be one LORD, Yehovah, and His name one.

This book, "The Malachi Matrix," has been written to bring glory to Yehovah's name and to encourage INDIVIDUALS to be reconciled to Yehovah, God the Father. The God of Abraham, Isaac, and Jacob. The most high God! The LORD, Yehovah of hosts. The one Jesus, Yeshua, called His Father and His God.

Malachi 3:16&17.

16. Then they that fear the LORD, Yehovah, spoke often one to another and the LORD, Yehovah, heard it and a book of remembrance was written before Him, for them that feared (reverenced) His name.

17. And they shall be Mine sayeth the LORD, Yehovah of host, in that day when I make up My jewels; I will spare them as a man spares his own children.

I hope you now see clearly that YEHOVAH, God the Father, is THE FOCAL POINT OF WORSHIP. I pray that you come to love and honor Yehovah's name.

Join my wife Jackie and I as we follow Yeshua, the Lamb of God, and worship the Father in spirit and in truth.

The Messenger of the Covenant

Malachi 3:6. Behold I will send My messenger and He shall prepare the way before Me, and the Lord (Adonai) whom you seek shall suddenly come to his temple, even the messenger of the covenant whom you delight in, behold He shall come, saith the LORD, Yehovah, of hosts.

As I outlined in chapter one of this book, the messenger that will prepare the way before Me, was John, the Baptist. The Lord whom you seek who shall suddenly come to his temple is Jesus, Yeshua. He is, without a doubt, "The Messenger of The Covenant". He is also The Mediator of a Better Covenant and The Mediator of The New Covenant.

Hebrews 9:14&15.

14. How much more shall the blood of Christ, who through the eternal spirit, offered Himself without spot to God, purge your conscience from dead works to serve the Living God.

15. and for this cause He is the Mediator of the New Covenant, that by means of death, for the redemption of the transgression that were under the first covenant, they which are called might receive the inheritance.

Hebrews 8.6. But now He has obtained a more excellent ministry, by how much also he is the Mediator of a Better Covenant.

Moses prophesied about this messenger of the covenant.

Deuteronomy 18:15&16.

15. The LORD, Yehovah, your God will raise up for you a prophet like me from your midst, from among your brethren. Him you shall hear

16. According to all you desired of the LORD, Yehovah, your God in Horeb in the assembly saying, let me not hear the voice of the LORD, Yehovah, my God, nor let me see this great fire anymore lest I die.

After Moses spoke these words, Yehovah, God the Father, responded by agreeing with what Moses had said and adding his own additional instructions.

Deuteronomy 18:17-18.

17. And the Lord Yehovah said to me: what they have spoken is good.

18. I will raise up for them a prophet like you from among their brethren and I will put My words in His mouth, and He shall speak to them all that I command Him.

19. And it shall be that whoever will not hear My words, which He speaks in My name, I will require it of him.

Do you realize what Yehovah is saying here? This one whom God the Father will raise up like Moses will speak the very words of God. He will come as a messenger from Yehovah, and notice brethren God the Father says: He shall speak IN MY NAME, In the name of and by the authority of Yehovah the most-high God.

Isn't that exactly what Jesus, Yeshua, said that He was sent as God's a messenger?

John 12:49&:50.

49. For I have not spoken of My own authority; but the Father who sent Me gave Me a command, what I should say and what I should speak.

50. And I know His commandment is eternal life. Therefore, whatever I speak just as the Father has told me, so I speak.

From past eternity Jesus, Yeshua, has been the obedient servant of Yehovah. He is The Messenger of the Covenant. He faithfully spoke his Father's words. Several times in the Old Testament, we find two God beings speaking directly to each other or about each other. Clear examples of this are found repeatedly in the book of Zechariah where Yehovah is speaking about Jesus, Yeshua.

Zechariah 13:7 Awake O sword against My Shepherd and against the man who is My fellow (companion) sayeth the LORD, Yehovah, Of hosts: Smite the Shepherd and the sheep shall be scattered, and I will turn my hand against the little ones.

My Shepherd can be none other than Jesus, Yeshua, the Christ.

The implications of this revelation are profound because most churches teach that the LORD of hosts spoken of in Zechariah is Jesus Christ. The LORD of host in Zechariah 13:7 and in Zechariah 14:16 are the same. This is none other than Yehovah, God the Father!

The most high God!

Zechariah 14:9 And the LORD, Yehovah shall be king over all the earth: in that day shall there be one LORD, and His name one.

This is the being that Jesus, "The Good Shepherd," referred to in John 20:17. Your Father and My Father, your God and My God.

This LORD of hosts is the same God who is spoken of in the book of Revelation, where He is often referred to as "LORD God Almighty."

Revelation 15:3&4.

3. And they sing the song of Moses the servant of God, and the song of the Lamb, (Stop) Notice two songs here, the song of Moses and the song of the Lamb. And who are they singing about?

3. (continued) saying, Great and marvelous are Thy works, LORD GOD ALMIGHTY, just and true are Thy ways Thou King of saints.

4. Who shall not fear Thee, O LORD, Yehovah, and glorify Thy name? for Thou art holy: for all nations shall come and worship before Thee: for Thy judgements are made manifest.

All nations shall come and worship before THE LORD GOD ALMIGHTY! God the Father! Exactly the same thing we find in Zechariah 14:6.

Zechariah 14:6 And it shall come to pass that every one that is left of all the nations which came against Jerusalem shall go up from year to year to worship the LORD, Yehovah, of hosts and to keep the Feast of Tabernacles.

The LORD, Yehovah of hosts, and the LORD God Almighty are the same God being, God the Father!

My whole point of showing these scriptures from Zechariah and Revelation is to prove the distinction between two God beings. Many have pushed the lie that Jesus, Yeshua, is Yehovah, the LORD of hosts mentioned in Zechariah. HE IS NOT! He is the forever faithful servant whom God the Father sent into the world as the Lamb of God. Yeshua did not send Himself! He is the Messenger

(Malek) of The Covenant. He is the Mediator of The New Covenant, A Better Covenant.

He is the one referred to as "My angel" In Exodus 23:23. Jesus, Yeshua, is called the angel of God in Exodus 14:19. He is the angel of the LORD in Zechariah 1:12&13.

12. Then the angel of the LORD answered and said O LORD of hosts, how long will Thou not have mercy on Jerusalem and on the cities of Judah against which Thou has had indignation these threescore and ten years?

13. And the LORD, Yehovah, answered the angel that talked to me with good and comforting words.

Here in the book of Zechariah, we have the angel of the LORD speaking to another being who is the LORD of hosts. Clearly, any honest student of the Bible would have to admit that the angel of the LORD here is the being who became Jesus, Yeshua, and the LORD, Yehovah, of hosts is God the Father. Jesus, Yeshua, IS NOT the LORD of hosts.

This has earth-shattering implications when Zechariah 14 is read with the indisputable truth that the LORD here is not Jesus, but YEHOVAH, God the Father.

Zechariah 14: 1 BEHOLD the day of the LORD, Yehovah, (God the Father) cometh and the spoil will be divided in the midst of thee.

Zechariah 14:3&4.

3. Then shall the LORD (God the Father) go forth and fight against those nations as He fought on the day of battle.

4. And His feet (God the Father's feet) shall stand that day on the mount of olives which is before Jerusalem...

God the Father. The Most High God, The LORD of hosts, The King over all the earth, is coming down to this earth! He will come to crush the rebellious nations, and He will make Christ's enemies His footstool.

Psalm 110:1. Yehovah said unto my Lord, Adonai, Sit at My right hand until I make Thy enemies Thy footstool.

Malachi 3:1 Behold, I will send My messenger and he shall prepare the way before Me: And the Lord, Adonai, whom you seek shall suddenly come to His temple, even the messenger {Malek) of the covenant whom you delight in, He shall come saith the LORD, Yehovah, of hosts.

Jesus, Yeshua, has been described by many titles that are unique to him alone. Emmanuel (God with us). The Good Shepherd. The Messiah (anointed one). Son of the living God. He has repeatedly and honorably called "The Lamb of God" in the Book of Revelation. The titles I will concentrate and focus on in the remainder of this chapter are below.

#1. The Messenger of The Covenant.

#2. The Mediator of a Better Covenant.

#3. The Mediator of The New Covenant.

Jesus, Yeshua, who was prophesied in Malachi 3:1 is specifically described as "the Lord (Adonai) whom you seek. "The Jews of Jesus" day were looking for a Messiah who they believed would come to liberate them from the Romans and re-establish the former glory like it was during King Solomon's reign.

Next, Malachi 3:1 state "He shall suddenly come to his temple". Jesus, Yeshua, burst upon the scene at the age of 30. Performing miracles, debating with the Scribes and Pharisees, and preaching in and around Jerusalem. He attracted large crowds everywhere He went, and the people delighted in his message.

The Messenger of the Covenant preached the Message of Reconciliation to Yehovah, God the Father. It is an invitation for Israel and the children of mankind from all nations to come home to the Creator God and enter into covenant with Him.

This message of RECONCILIATION Was also prophesied in the book of Ezekiel. This invitation to come back home is personal. Jesus, Yeshua is the Good Shepherd who is leading INDIVIDUALS back to the Father, teaching us to embrace the true worship of Yehovah. He exampled us in every aspect of loving one another. It is a matter of the heart and is a matter of INDIVIDUAL TRANSFORMATION!

Ezekiel 11:19&20

19. And I will give them one heart, and I will put a new spirit within you: and I will take the Stony heart out of their body and give them a heart of flesh.

20. That they may walk in my statutes and keep my ordinances and do them; and they shall be my people, and I will be their God.

It is indeed a matter of the heart and the Transformation of Individual Hearts so what more shall we say of this matter about the ministering of the new covenant by The Messenger and Mediator of the same?

Hebrews 12:18 and 22-24.

18. For you are not come unto the mount that might be touched, and that burned with fire, nor to blackness and darkness and Tempest.

22. But you are now come unto Mount Zion, and unto the city of the Living God, the heavenly Jerusalem into an innumerable company of angels.

23. To the General Assembly of the church (ecclesia) of the firstborn, which are written in heaven, and to God the judge of all, and to the spirits of just men made perfect.

24. And to Jesus, Yeshua, the Mediator of The New Covenant, and to the blood of sprinkling, which speaks better things than that of Abel.

The establishment of the new covenant was of paramount importance in the heart and mind of Jesus, Yeshua. You cannot live in covenant with God without accepting and entering into this NEW COVENANT, which Jesus, Yeshua, established on better promises. So, what exactly is this New Covenant?

Hebrews 8:6-8 and 10.

6. But now He, Yeshua, has obtained a more excellent ministry by how much also he is The Mediator of a Better Covenant, which was established on better promises.

7. For if the first covenant was faultless, then should no place have been found for a second.

8. But finding fault with them He sayeth, behold the days come sayeth the LORD, Yehovah, when I will make a New Covenant with the house of Israel and the house of Judah.

10. For this is the covenant that I will make with them after those days sayeth the LORD, Yehovah: I will put my laws into their minds, and write them on their hearts, and I will be to them a God, and they shall be my people.

Remember, brethren, that salvation is a matter of the heart. True worship of Yehovah, God the Father, is also a matter of the heart, just as Jesus and Yeshua taught us.

Matthew 22:37. Jesus Yeshua said to him, thou shalt love the LORD, Yehovah, with all your heart, and with all your soul, and with all your mind.

As I stated earlier in this chapter, Jesus, Yeshua, is the Malik (messenger) of the New Covenant. He established The New Covenant with his disciples on the night he was betrayed. All those whom Yehovah, God the Father, is calling will recognize the historic and eternal significance of this New Covenant Memorial. You will notice that I do not and will not refer to it as "The Passover." The word Passover was never used or spoken until the 1500s. William Tyndale coined the word in 1536. In his sincere attempt to translate the Bible into English, he made up or coined "Passover." In examining the biblical spring festival, Tyndale observed that two ancient Hebrew words were pronounced the same - the verb "pasach" meaning to skip or pass over, and the noun "pesach" referring to the lamb. He translated them both "Passover" causing a plethora of confusion that exists to this day.

When Jesus, Yeshua, sent Peter and John to go and prepare the Passover that we may eat, He actually sent them to go prepare the "pesach" (the lamb)

Luke 22:8 and he sent Peter and John saying go and prepare us the Passover (the lamb) that we may eat.

Dealing with this issue, let me state that I believe that Jesus, Yeshua, did everything and gave every command with a sense of purpose. He spoke every word intentionally to facilitate and accomplish the plan of Yehovah, God the Father. When we fully understand that Jesus specifically told Peter and John to prepare "the Lamb" so that we may eat, the events of that night become much clearer. Notice that Jesus, Yeshua, made it clear that he would indeed eat the lamb with his disciples.

Luke 22:14&15.

14. And when the hour had come, He sat down, and the twelve disciples with Him.

15. And he said unto them, with desire I have desire to eat this Passover (this lamb) With you before I suffer.

So then, almost the whole Christian world somehow concludes from this that they never ate the Passover? Do you see how faulty this line of thought is? Do you actually believe that Jesus, Yeshua, gave a command to prepare the lamb and that he stated His heart's desire was to eat the lamb with his disciples, and then he never ate the Passover lamb?

I'm not alone in this conclusion. The late Mr. Ronald L. Dart of Christian Education Ministries believed that Christ and His disciples actually ate a lamb that night. The parent Church of God, 7th Day, which Herbert Armstrong came out of, has a study titled: The Lord's Supper. Though I don't agree with all their conclusions, I do believe they made some true and profound statements on this topic. Opening statement: The Lord's Supper Service is a solemn Christian Memorial of Jesus's crucifixion. Introduced on the night he was betrayed, it commemorates his death on Calvary's cross and as a sacrifice for our sins that we might live.

Under Origins of the Lord's Supper Ordinance, we find this,

Matthew, Mark, and Luke report that as the time of the annual observance of the Passover approached, Jesus instructed Peter and John to make the necessary preparations for him and the 12 disciples to eat together that evening. As Jesus and the disciples reclined around the table together, eating what would be their last Passover meal together, Jesus did something different, something that was to be practiced by his followers for all time as a memorial of His Death.

Luke 22:19. And He took bread, gave thanks and broke it, and gave it to them saying:

This is my body given for you; do this in remembrance of Me.

Matthew 26:27828.

27. And he took the cup and gave thanks and gave it to them saying, drink ye all of it.

28. For this is my blood of The New Covenant, which is poured out for many for forgiveness of sins.

Understand, brethren, what I am proposing to you. Jesus, Yeshua, as The Mediator of a Better Covenant, as the Mediator of The New Covenant, established THE NEW COVENANT MEMORIAL! If we're going to use a term, let us use the terms Jesus, Yeshua used. He said distinctly, "do this in remembrance of Me." A MEMORIAL! He also said clearly, "this is the blood of The New Covenant"! Jesus, Yeshua, called the observance He instituted on the night He was betrayed THE NEW COVENANT MEMORIAL!

Now, I can hear the detractors screaming. No Way! They could not have eaten a Passover meal, a lamb, that night because the priest didn't begin sacrificing the lambs until the following evening. I have done some intense research trying to answer this question. On a site titled (Bible Tools) under 'What the Bible says about Passover Kept at the Temple,' I found this:

While God intended the Passover (Chag Ha-Pesach) and the feast of unleavened bread (Chag Ha-Matzot) to be separate, though distinct, observances, the Jews ended up combining the two during the Babylonian exile, as the encyclopedia Judaica confirms: the feast of Passover consists of two parts: the Passover ceremony and the feast of unleavened bread. Originally, both parts existed separately, but at the beginning of the Babylonian exile, they were combined.

Volume 13p. 139. This careless and unscriptural merging of festivals resulted in the Jews observing Passover late on Abib 14, just hours before the feast of unleavened bread began.

At the time of Jesus Christ, this mixture was on full display. Philo of Alexandria in Divita Mosis notes that in the early 1st century, the Passover was not strictly a temple-kept event but was one in which people also killed their own lambs without the help of the priest. In his writings, War Of The Jews, Flavius Josephus records that in four BC, over 250,000 lambs were sacrificed for Passover. However, given the limited space in the temple environs and the fact that Jewish tradition (not the word of God) held that the lambs were to be slain within a two-hour time slot (from the ninth to the eleventh hour or 3:00-5:00 pm) it is readily apparent that not all these lambs could have been sacrificed at the temple. In fact, Joachim Jeremias, who wrote 'Jerusalem In The Time Of Christ,' calculates that the three courses (shifts) of priests on duty could slay only 18,000 lambs during those two hours. Josephus records that the rest of the lambs, a far greater number, were slain in their own homes.

Another critical point is that despite Passover and unleavened bread being distinct festivals, they were commonly grouped together and simply called Passover. Thus, when the gospel writers mentioned Passover, it can sometimes refer to the Passover sacrifice itself (Matthew 26:17 Mark 14:12), the day the sacrifice was made (Mark 14:1), or the whole eight-day period of Abib, Passover plus Unleavened Bread, Luke 22:1.

In actuality, then, there were really two Passover observances happening at the time of Christ. One was led by the priest at the temple, and another was observed by the people in their homes. These separate observances were also at different times. The temple kept Passover was observed late in the afternoon of Abib 14, while the home kept Passover was kept at the beginning of Abib.

14. As the gospels show, Jesus, Yeshua, and his disciples ate the Passover in a home, rather than at the temple, observing it the evening before the priest did in the temple. In other words, Jesus, Yeshua, kept it as Abib 14 began, while the priests kept it as Abib 14 ended.

(STOP) Written by David C. Grabbe.

My whole point in sharing this detailed historical information is to drive home the fact that Jesus, Yeshua, desired to and, in fact, did eat a Pesach lamb with his disciples, thus fulfilling his Father's command to keep the Chag Ha- Pesach.

To those who want to claim that this was just an ordinary meal, I would ask them: So, when did the disciples keep the feast of Passover? The Chag Ha- Pesach? They were Torah observant Jews. They had kept or observed the Passover (Chag-Ha Pesach) all their lives. There is no record of them observing it the next evening at the temple with the priests. There is no record of them keeping it a month later.

The following facts are indisputable. Jesus, Yeshua told Peter and John to go prepare the Passover (The Chag Ha-Pesach), which is a lamb. Jesus said: With desire, I have desired to cat this Chag-Ha-Pesach (this lamb) with you before I suffer. So then, while they were gathered to eat the Pesach meal, what happened?

Mark 14:22-25.

22. And as they were eating, He took bread, and after blessing it broke it and gave it to them, and said: take this is my body.

23. And He took the cup, and when He had given thanks he gave it to them; And they drank all of it.

24. And He said unto them: this is the Blood Of The Covenant, which is poured out for many.

Jesus, Yeshua, as the Mediator of the New Covenant was introducing the symbols of the covenant that he established as the MEMORIAL OF HIS DEATH and the eternal Truth that he is The Lamb of God who takes away the sins of the world.

I Corinthians 11:23-26.

23. And He took the cup, and when He had given thanks, he gave it to them; And they drank all of it.

24. And He said unto them: this is the Blood Of The Covenant, which is poured out for many.

Jesus, Yeshua, as the Mediator of the New Covenant was introducing the symbols of the covenant that he established as the MEMORIAL OF HIS DEATH and the eternal Truth that he is The Lamb of God who takes away the sins of the world.

I Corinthians 11:23-26.

23. For I have received of the Lord that which also I delivered unto you, that the Lord Jesus, Yeshua, the same night in which he was betrayed took bread:

24. And when he had given thanks, he broke it, and said, take eat; this is My body which is broken for you: this Do IN REMEMBERANCE OF ME.

25. After the same manner also he took the cup, when he had supped, saying this cup is THE NEW COVENANT IN MY BLOOD, this do as often as you drink it, IN REMEMBERANCE OF ME.

26. For as often as you eat this bread, and drink this cup, you do show the Lord's death till He comes.

Notice that Jesus, Yeshua, did not call what he was doing "The New Covenant Passover." He specifically called it THE NEW COVENANT, and twice He said: this Do IN REMEMBERANCE OF ME!

The word remembrance here is from the Greek word "anamnesis" Strongs #364 from the root word (anamimnesko, bring to mind) definition: properly, deliberate, recollection, done to better appreciate the effects (intended results) of what happened; active self-prompted recollection, especially as a memorial. (memorial sacrifice).

So, when Jesus, Yeshua said: "This do in remembrance of Me," he was saying this is a memorial so that you never forget. This is why, brethren, I will not call this observance The New Covenant Passover. Moving forward, out of respect for Yeshua's own words, I will call it THE NEW COVENANT MEMORIAL!

Sadly, there are some who, in their zeal to get rid of the ecclesia of any remnants of false teachings, say that what took place that night was Jesus, Yeshua, betrayed was just an ordinary meal. To quote one advocate of this idea: "Jesus would not have changed the date of the Passover." There are some who no longer observe the ceremony commonly referred to as 'The New Covenant Passover.' They also no longer observe the foot washing, which this author believes Jesus, Yeshua, introduced as a Service of Humility.

Jesus, Yeshua, didn't change a holy day. He introduced something Completely New.

THE NEW COVENANT MEMORIAL. As the messenger of The Covenant and as the Son of the Living God, He had full authority and responsibility to establish this memorial. How pray tell, are we to be able ministers of the new covenant if we don't partake of the New Covenant Memorial?

II Corinthians 3:3-6

3. For as much as you are manifestly declared to be the epistles of Christ ministered by us, written not with ink, but with the spirit of the Living God; not in tables of stone, but in the fleshly tables of the heart.

4. And such trust have we through Christ to God-ward (directed toward God).

5. Who has also made us Able Ministers of the New Covenant; not of the letter, but of the spirit, for the letter killeth, but the spirit giveth life.

Once again, I declare unto you, brethren, you cannot remain in covenant with Yehovah, God the Father, if you refuse to participate in The New Covenant Memorial, which was instituted by Yeshua, the Son of the Living God.

John 6:53-54

53. Then Jesus, Yeshua, said unto them, verily, verily I say unto you, except you eat the flesh of the son of man, and drink his blood, you have no life in you.

54. Whosoever eateth My flesh, and drinketh My blood, hath eternal life: and I will raise him up at the last day.

Jesus, Yeshua is THE MEDIATOR OF THE NEW COVENANT!

Jesus, Yeshua, is THE MESSENGER OF THE COVENANT!

The Identity and Message of The Two Witnesses

Revelation 11:3&4.

3. And I will give power to My two witnesses, and they shall prophecy a thousand two hundred and threescore days clothed in sackcloth.

4. These are the two olive trees and the two candlesticks standing before the God of the whole earth.

Zechariah 4:11&14.

11. Then answered I, and said unto him, what be these two olive trees upon the right side of the candlesticks and upon the left side thereof?

14. Then said he, These are the anointed ones who stand by the Lord of the whole earth. There is a biblical principle that has been commonly accepted throughout history among God's people.

Deuteronomy 19:15. One witness shall not rise against a man concerning any iniquity or sin that he commits; by the mouth of two or three witnesses, the matter shall be established.

Matthew 18:16. But if he will not hear, take with you one or two more, so that by the mouth of two or three witnesses every word may be established.

Hebrews 10:28. Anyone who has rejected Moses' law died without mercy on the testimony of two or three witnesses.

It is amazing how often in the Bible, Yehovah God and Jesus, Yeshua, commissioned, commanded, and sent two witnesses to deliver a message of warning and a message of reconciliation. The two witnesses of the book of Revelation will also deliver a message of warning and reconciliation. I find it interesting that the last book of the Old Testament and the last book of the New Testament both reference two such witnesses.

Malachi 4:4-6

4. Remember the law of Moses My servant, which I commanded unto him in Horeb for all Israel, with the statutes and judgements.

5. Behold I will send you Elijah the prophet before the coming of the great and dreadful day of the LORD, Yehovah.

6. And he shall turn the heart of the fathers to the children and the hearts of the children to the fathers, lest I come and smite the earth with a curse (total destruction).

This book, The Malachi Matrix, is offered as an archetype declaring and detailing the reoccurring cycle of people turning away from God and embracing idolatry and pagan forms of worship. Yehovah has repeatedly sent His prophets, men like Moses and Elijah, with a warning and a message of reconciliation.

Revelation 11:3&4

3. And I will give power unto My two witnesses, and they shall prophecy one thousand two hundred and threescore days, clothed in sackcloth.

4. These are the two olive trees and the two candlesticks standing before the God of the whole earth.

Two individual men of God are mentioned in the last chapter of Malachi who will bring a witness of warning and reconciliation to a dying world before the coming of the great and dreadful day of the LORD, Yehovah. These two men are Moses and Elijah. This prophecy was not fulfilled in Malachi's day. It was not completely fulfilled in Christ's first coming, except in the picture of John the Baptist being a type of Elijah. It will be ultimately fulfilled in the not-too- distant future, just prior to the coming of Yehovah, God the Father, and Jesus, Yeshua.

Why Moses and Elijah?

Moses is the great law giver of Exodus.

Elijah is the great archetype of all the prophets. They're both together the great defenders of the faith.

Do I believe, as some Protestant churches have proposed that these two witnesses will be the actual persons of Moses and Elijah, who have been alive up in heaven these thousands of years? Absolutely not! I have heard it preached that Moses went alone up into a high mountain, and no one actually saw him die, so God somehow could have preserved his physical life up in heaven. More sermons have contended that Elijah was raptured away to heaven by Chariots of fire, and he was never seen or heard from on earth again. This unscriptural notion also insists that Elijah never died and is still alive up in heaven. Concerning the first myth about Moses, the scriptures clearly say that Moses died.

Deuteronomy 34:5&6

5. So Moses the servant of the LORD, Yehovah, died there in the land of Moab, according to the word of the LORD, Yehovah.

6. And they buried him in a valley in the land of Moab, over against Beth peor: no man knoweth his sepulcher unto this day.

Concerning the second theory that Elijah was taken physically to heaven, which is the domain of Yehovah, God the Father, and all the holy angels, I simply say NO WAY! The Bible references 3 heavens. The heaven, which is the dwelling place of Yehovah, God the Father. The heavens, which is the great expanse of the universe and all the stars. And finally, heaven, which is the sky or atmosphere that surrounds our planet earth. Elijah was carried away up into the sky someplace on earth because it was time for him to retire and allow his apprentice, Elisha, to take up his mantle in service to Yehovah.

There is proof that Elijah was not physically transported to God's heavenly throne because later in the historical time sequence, Jehoram, King of Judah, received a letter from Elijah, who was obviously still alive somewhere on Earth. This King Jehoram of Judah was the Son of Jehoshaphat. Jehoshaphat reigned in Judah during the same time that the wicked King Ahab was ruling in Israel. These kingdoms were two separate nations at this time. Ahab was the King of Israel during Elijah's active ministry.

II Chronicles 21:12-15

12. And there came a writing (letter) from Elijah the prophet saying: thus, sayeth the Lord, Yehovah, God of David thy Father, because thou hast not walked in the ways of Jehoshaphat thy Father, nor in the ways of Asa king of Judah.

13. But has walked in the ways of the kings of Israel, and has made Judah and the inhabitants of Jerusalem go a whoring, like to the whoredoms of the House of Ahab, and has slain thy brethren of thy Father's house, which were better than thyself:

14. Behold with a great plaque will the LORD, Yehovah, smite thy people and thy children, and thy wives, and all thy goods.

15. And thou shalt have great sickness by disease of the bowels, until thy bowels fall out by reason of the sickness day by day.

So, then brethren, if Elijah was alive in the flesh in the very throne room of Yehovah God up in heaven, that would be some expensive postage to mail a letter from heaven down to King Jehoram. The simple truth is that Elijah was raptured up by the chariots of fire into the sky and taken to an undisclosed location on this Earth to live out his life in retirement.

Having said all this, I will allow that if Yehovah, the God of Creation, wants to resurrect Moses and Elijah from the dead. He is capable of doing exactly that. I believe it is far more likely that the two witnesses of Revelation chapter 11 will be two devout individuals born in our age who will each have the characteristics of Moses and Elijah. It is apparent from scripture that these two will be endowed with miraculous power directly from the dynamic spirit of Yehovah, God the Father. Miraculous power will be one of the signs proving they have been sent by Yehovah.

Revelation 11:3-6

3. and I will give power unto My two witnesses, and they shall prophecy a thousand two hundred and threescore days clothed in sackcloth.

4. These are the two olive trees and the two candlesticks standing before the God of the whole earth.

5. And if any man will harm hurt them fire proceeds out of their mouth, and devours their enemies: and if any man will hurt them, he must in this manner be killed.

When it says here in verse 5, "fire proceeds out of their mouth," it's not like they are fire-breathing dragons. I believe just like Elijah did in 2nd Kings chapter 1; they speak the words and fire falls from heaven, destroying their enemies.

Notice, also in verse 3 it says, they are clothed in sackcloth. The will not be your slick T.V evangelists with fancy three-piece suits and silk ties. They will be clothed much like Elijah and John the Baptist. Dressed in casual attire, perhaps blue jeans and possibly leather jackets. I have a picture in my mind of how Moses may have dressed. It was likely the humble appearance of a shepherd. Moses had herded sheep for 40 years in the land of Midian for His Father-in-law Jethro before returning to Egypt.

Jesus, Yeshua, did speak about John the Baptist's attire.

Matthew 11:8 But what went you out to see? A man clothed in soft raiment? Behold, they that wear soft clothing live in King's houses.

John the Baptist was a type of Elijah, so their dress was probably similar.

King Ahaziah, who was the wicked son of King Ahab, once enquired of some of his messengers about a prophet who had spoken to them.

11 Kings 1:7&8

7. And he (King Ahaziah) said unto them, what manner of man was he which came up to meet you, and told you these words?

8. And they answered him. He was a hairy man, and gird with a girdle of leather about his loins. And answered and said, it is Elijah the Tishbite.

Moses was raised in the House of Pharoah, King of Egypt. He was, according to Jewish tradition, trained as a military commander of Pharoah's army. He was by all accounts a goodly man, meaning he was likely tall in stature. After Moses had several close encounters with Yehovah, the God of Israel, he was endowed with not only a glow about him but also a physical strength and endurance that lasted his entire life.

Deuteronomy 34:7

Moses was 120 years old when he died. His eyes were not dim nor his natural vigor diminished.

Returning now to the prophet Elijah and hints of his physical prowess, we have this astounding account

I Kings 18:44-46

44. And it came to pass at the 7th time that he (Elijah) said, behold there ariseth a little cloud out of the sea, like a man's hand. And he said, go up and say unto Ahab, prepare thy chariot, and get thee down, that the rain stopped thee not.

45. And it came to pass in the meanwhile, that the heaven was black with clouds, and wind, and there was a great rain. And Ahab rode and went to Zezreel.

46. And the hand of the LORD, Yehovah, was on Elijah, and he girded up his loins and ran before Ahab to the entrance of Zezreel.

WOW! What kind of physical prowess did Yehovah bestow upon Elijah so they could outrun A chariot? I hope by now you're getting a picture of what kind of man these two witnesses of revelation will be. They will be strong and powerfully built men. I can picture them both roaring into Jerusalem on Harley-Davidson motorcycles dressed in leather. They will be Hebrews in the tradition of the Maccabees. Side by side, these two Sons of Thunder will command attention on the world stage through their stinging rebuke against the beast power and by the miraculous plagues they pronounce upon a lawless people.

I believe that another unmistakable sign of their bona fides is that they will refer to God the Father by His revealed and eternal name, YEHOVAH! They will also, I believe, refer to Jesus as

YESHUA! Moses is the faithful servant of God who was chosen as the human being who He would reveal His name to.

Exodus 3:15 and God said moreover (furthermore) unto Moses, thus shall you say unto the children of Israel, the LORD, Yehovah, God of Abraham, and the God of Isaac, and the God of Jacob hath sent me unto you: this is My Name Forever, and My Memorial unto all generations.

Before this momentous event, the believers in the creator God only knew Him by His titles.

Exodus 6:2-3

2. And God spoke unto Moses and said unto him, I am the LORD, Yehovah.

3. And I appeared under Abraham, unto Isaac and unto Jacob by the name of God Almighty (El-Shadi), but by My name YEHOVAH was I not known unto them.

When Moses first approached Pharaoh, the King of Egypt, he used God's proper name, Yehovah.

Exodus 5:1-3

1. And afterwards Moses and Aaron went in and told Pharaoh, Thus sayeth the LORD, Yehovah, God of Israel, let My people go, that they may hold a feast unto Me in the wilderness.

2. And Pharaoh said, who is Yehovah that I should obey His voice to let Israel go? I know not Yehovah, neither will I let Israel go.

3. And they said, the God of the Hebrews hath met with us: let us go, we pray thee, three days journey into the desert and sacrifice unto Yehovah our God.

Notice! Moses and Aaron said to Pharaoh, King of Egypt, Yehovah says, Let My people go! To which Pharaoh answers: Who is Yehovah? I don't know no Yehovah!

Moses was intimately aware of the name of the God who had spoken to him from the burning Bush. YEHOVAH is the proper name of the God of Israel. The very first thing this same God did when he introduced himself to the children of Israel at Mount Sinai was to tell them his awesome name.

Exodus 20:1&2

1. and God spoke all these words saying,

2. I am the LORD, YEHOVAH thy God, which brought thee out of the land of Egypt, out of the House of bondage.

One of the witnesses from the Book of Revelation who is the resurrected Moses or a man who has the spirit of Moses, will also be intimately familiar with the name of God. Mark my words! He will boldly declare the awesome name of YEHOVAH to the whole world.

The second man of the two witnesses will either be the resurrected Elijah, the prophet, or a man who has all of Elijah's attributes and character. The prophet Elijah, who had the showdown with Ahab and the prophets of Ba-el was also unashamed of Yehovah's name. He used Yehovah's name in everything he said and did.

The northern Kingdom of Israel had been bewitched and corrupted by Jezebel, the evil Zidonian wife of Ahab. Together, they had promoted the idolatrous worship of Ba-el. The people of Israel had mixed this Pagan worship with the worship of the God of Israel. Ba-el worship involved bowing to the images of the golden calf and the apus bull. They danced around these idols participating in drunken orgies. They sacrificed their children, burning them to death in the fire. They performed grisly torture involving self-mutilation and bloodletting. Yehovah God finally had enough of this putrid mixing of unclean and satanic worship. He sent Elijah, the prophet, to punish Israel with a drought just to get their attention.

Elijah was an imposing figure in his own right, and he was a man of few words. When he did speak, he seldom failed to use Yehovah's name.

I Kings 17:1 and Elijah the Tishbite, who was of the inhabitants of Gilead, said unto Ahab, As Yehovah, God of Israel lives, before whom I stand, there shall not be dew or rain these years, but according to my word.

Then Elijah simply turned around and left. Ahab didn't see or hear from him for three years. After many days, Yehovah told Elijah, "It is enough, I will send rain upon the earth."

Ahab had been searching everywhere for Elijah to kill him with no success. Elijah had been hiding at Zarephath, which was the capital city of the Zidonians. He was being hidden and cared for by a poor Zidonian widow. Elijah finally met with Obadiah, who was the governor of Ahab's house. Now, this Obadiah was a man who feared Yehovah greatly. Elijah told Obadiah to go tell Ahab to meet him on a high mountain. He also instructed him to bring his 450 Prophets of Ba-el for a showdown.

1 Kings 18:16-24

16. So Obadiah went to meet Ahab and told him; and Ahab went to meet Elijah.

17. And it came to pass, when Ahab saw Elijah, that Ahab said unto him, art thou he who troubles Israel?

18. And he answered, I have not troubled Israel: but you and your Father's house, in that you have forsaken the commandments of Yehovah, and have followed Ba-el.

19. Now therefore send, and gather to me all Israel unto Mount Carmel, and the prophets of Ba-el 450, and the prophets of the Groves 400, which eat at Jezebel's table.

Elijah seemed to think that 450 to one was about fair.

20. So Ahab sent unto all the children of Israel and gathered the prophets together unto Mount Carmel.

21. And Elijah came unto all the people and said, how long halt you between two opinions? If Yehovah be God follow him, and if Ba-el then follow him. And the people answered him not a word.

22. Then said Elijah unto the people, I even I only remain a prophet of Yehovah, but Ba- el's prophets are 450 men.

23. Let them therefore give us two bullocks: and let them choose one Bullock for themselves, and cut it in pieces and lay it on the wood, and put no fire under.

24. And call you on the name of your gods: and I will call on the name of YEHOVAH and the God who answers by fire, let him be God. And all the people answered and said, it is well spoken.

25. And Elijah said unto the prophets of Ba-el, choose you one Bullock for yourself, and dress it first; for you are many; and call upon the name of your gods, but put no fire under it.

26. And they took the Bullock which was given them, and dressed it, and called on the name of Ba-el from morning even until noon, saying O Ba-el hear us. But there was no voice, nor any that answered. And that leaped upon the altar which was made.

27. It came to pass at noon, that Elijah mocked them, and said, cry aloud; for he is a god: either he is on a journey, or peradventure he sleepeth and must be awakened.

Now, that's not very nice, making fun of someone else's god or their religion, but that's exactly what Elijah did. We have this idea that permeates churches and much of society today that promotes the nonsense that all religions are equal and just different paths to God or to heaven. I would like to go on record here now in stating that this idea is a lie straight from Satan himself. Jesus, Yeshua, said as much. John 14:6. Jesus, Yeshua, sayeth unto him, I am the way, the truth, and the life: no man cometh unto the Father, but by Me.

There are not multiple ways to be saved or multiple paths to heaven. Anyone who tells you otherwise is a liar, and the truth is not in him.

(back to our story)

28. And they cried aloud, and cut themselves after their manner with lancets, till the blood gushed out upon them.

29. And it came to pass, when midday was passed, and they prophesied until the time of the offering of the evening sacrifice, that there was neither voice, nor any to answer.

By this time the prophets of Ba-el had probably collapsed from exhaustion. It is interesting to me that they thought all their crazed gyrations would produce some supernatural results. It is possible, even likely, that they had conjured up some miraculous, demonic, and satanic wonders in the past, but not this time. Yehovah, God made sure of it.

Now, it was Elijah's turn. For the sake of time, I will paraphrase. Elijah repaired the altar of Yehovah that was broken down. Used 12 Stones which represented the 12 tribes of Israel. He had them cut up the Bullock allotted to him. In verse 31 it mentions that he built an altar in the name of (dedicated to) YEHOVAH! He dug a trench all around the altar and he had them fill four barrels with water and poured on the sacrifice and on the wood. He told them to do it a second time, he told them to do it a third time, and they did it a third time, and the water filled the trench all around. I guess they wanted to prove to them that this isn't a magic trick. At this point, we will return to the scripture for the climax of this historic showdown.

1 Kings 18:36-40

36. And it came to pass at the time of the offering of the evening sacrifice that Elijah the prophet came near and said, YEHOVAH, God of Abraham, Isaac, and Jacob, let it be known this day that you are God in Israel and that I have done all these things at Thy word.

37. Hear me O YEHOVAH, hear me, that this people may know that thou art God and that thou hast turned their heart back again.

38. Then the fire of YEHOVAH fell and consumed the burnt sacrifice and the wood, and the stones, and the dust, and licked up the water that was in the trench

39. And when the people saw it, they fell on their faces and they said,

YEHOVAH He is God!

YEHOVAH He is God!

40. And Elijah said unto them, take the prophets of Ba-el: let none of them escape. And they brought them down to the brook Kishon and slew them there.

Once again, what kind of man was Elijah that he slaughtered 400 prophets of Ba-el assembly line style, one after the other with a sword?

The point of this whole story is to once again drive home the fact that the resurrected Elijah, or a man with the spirit of Elijah, will be one of the two witnesses of revelation. Just like Moses, he will be personally and intimately convinced that God the Father's name is YEHOVAH! He will be about the business of turning the hearts of the fathers to the children and the hearts of the children to the fathers. The Ministry of RECONCILIATION! The reconciliation of families based on the biblical model and, more importantly, the reconciliation of mankind to YEHOVAH, God the Father.

At this juncture, I want to point out two truths from the Book of Revelation about the importance of the Father's name, which is YEHOVAH.

It's commonly accepted that the era of the Philadelphia church, or the spirit of the Philadelphia church, will be present in the last days. Hear now, my brother, what the Angel (the messenger) has to say to the ecclesia.

Revelation 3:7-9.

7. Unto the Angel of the church in Philadelphia write; these things sayeth He that is holy, He that is true, He that has the key of David, He that openeth and no man shutteth and shutteth, and no man opens.

8. I know that works, behold I have set before thee an open door, and no man can shut it; for thou hast a little strength, and has kept My word, and HAS NOT DENIED MY NAME!

9. Behold, I'll make them of the synagogue of Satan, which say they are Jews, and are not, but do lie; them I will make them to come and worship before thy feet, and to know that I have loved you.

Verse 12. Him that overcometh will I make a pillar in the temple of my God and he shall go no more out; and I will write upon him THE NAME OF MY GOD, and the name of the city of my God which is NEW JERUSALEM, which comes down from heaven from my God, and will write on him My new name.

In the last days, just before the coming of the great and dreadful day of Yehovah, it will be made evident to all men who dwell on the face of the earth that they must choose! The choice will be between worshipping the man of sin, the lawless one, who is the beast, or worshiping YEHOVAH, God the Father.

II Thessalonians 2:3&4

3. Let no man deceive you by any means, for that day shall not come, except there come a falling away first, and that man of sin be revealed, the son of perdition.

4. Who opposes and exalteth himself above all that is called God, or that is worshipped: so he as God sitteth in the temple of God showing himself that he is God.

Revelation 14:9&10

9. And the third Angel followed them saying with a loud voice, if any man worship the beast and his image, and receive his mark in his forehead or in his hand.

10. The same shall drink of the wine of the wrath of God which is poured out without mixture into the cup of His indignation, they shall be tormented with fire and brimstone in the presence of the holy angels, and in the presence of the Lamb.

The two witnesses will oppose this man of sin, who will demand that all men worship him as a God! The two witnesses will speak YEHOVAH'S name plainly to the whole world. They will also plead with all men to turn their hearts back to Yehovah, God the Father.

Their message will be the same message written in the book of Malachi, RECONCILIATION! Come home to Yehovah, God the Father.

Those who heed the invitation will be marked with God the Father's name, Yehovah written on their foreheads.

Revelation 14:1 I looked and lo, a LAMB stood on Mount Zion and with him a hundred and forty-four thousand, having HIS FATHER'S NAME written on their foreheads.

One witness will be the resurrected Moses, or man with the spirit of Moses. Moses represents the law of Yehovah, the 10 commandments. The beast of revelation will trample underfoot the law of God, so naturally, there will be a great confrontation.

The second witness will be the resurrected Elijah or a man with the spirit of Elijah. Elijah represents the prophets of Yehovah. All the prophets down through the ages have preached a message of warning and a message of RECONCILIATION to Yehovah, God the Father.

Once again, there will be a great showdown between the forces of good, the sons of light, and the forces of evil, the sons of darkness.

Revelation 13:15-17

13. And he had power to give life unto the image of the beast, that the image of the beast should speak and cause as many as would not worship the image of the beast should be killed.

16. And he causes all, both small and great, rich and poor, free and bond, to receive a mark in their right hand or in their foreheads.

17. And no man might buy or sell, save he had the mark of the beast or the number of his name.

These two witnesses of revelation will proclaim to all men: it is time to choose! How long will you halt between two opinions? If Yehovah be God, follow him, but of the beast, follow him! How will you choose, dear friend? It is a matter of Life or Death! It is my heart's deepest hope and my fervent prayer that Yehovah, God the Father awakens within your spirit a desire to repent of sin and that He turns your heart back to Him.

Jesus, Yeshua is the Messenger of the New Covenant. He is referred to repeatedly in the Book of Revelation as THE LAMB OF GOD! He offered Himself as the spotless Lamb who died as a sacrifice for all my sins and all of your sins.

Matthew 26:28 For this is My blood of the NEW COVENANT which is shed for many for the remission (forgiveness) of sins.

If your spirit has been stirred by the message in this book, do not harden your heart. Listen to The Messiah's words!

Matthew 11:25-29

25. At that time Jesus, Yeshua, answered and said, I thank Thee O Father, Lord of heaven and earth, because Thou hast hid these things from the wise and prudent, and has revealed them unto babes.

26. Even so Father, for it seemed good in Thy sight.

27. All things are delivered unto Me, of My Father and no man knows the Son but the Father, neither knoweth any man the Father, save (without) the Son, and he to whom the Son will reveal Him.

28. Come unto Me, all you that labor and are heavy laden, and I will give you rest.

29. Take my yoke upon you and learn of Me: for I am meek and lowly in heart: and you shall find rest for your souls.

30. For My yoke is easy and My burden is light.

The invitation is extended to every INDIVIDUAL, and it is Personal! You cannot come to know God the Father without the Son, the Lamb of God who takes away the sins of the world. The question, my friends is this: who will you choose to follow, and who will you worship?

If Yehovah be God, follow Him!

If the beast be God, follow him!

Oh, how I pray and long for the day when the people of God will answer with one accord.

YEHOVAH HE IS GOD!

YEHOVAH HE IS GOD!

Homecoming/ Worshipping the Father in Spirit and in Truth

1 Corinthians 8:6 But to us there is but one God, the Father, of whom are all things, and we in Him: and one Lord (master) Jesus Christ, by whom are all things, and we in Him.

Matthew 20:25-27

25. But Jesus, Yeshua, called them to himself and said, You know that the princes of the Gentiles exercise dominion over them, and they that are great exercise authority upon them.

26. But it shall not be so among you: but whoever will be great among you: let him be your minister.

27. And whosoever will be chief among you, let him be your servant.

If you have been affiliated with any of the Churches of God over the last 50 years, as I have been, you must admit that there has been a reoccurring problem with men who, to varying degrees, have lorded it over God's people. This cancer of authoritarian leadership is prevalent in almost all the different denominations that call themselves Christian.

You name it and take your pick: Roman Catholic, Lutheran, Protestant, Church of England, Eastern Orthodox, Methodist, Worldwide Church of God, and all the branches thereof, Mormon, Amish, or Rock and Roll Evangelical.

All the way from the Pope (Roman Catholic). "The Apostle" (Worldwide Church of God). Patriarch of Constantinople (Eastern Orthodox). Apostle (Mormon). Bishop (Methodist and Amish).

President and National Director (Vineyard Churches). Supreme Governor and Archbishop (The Reigning King or Queen of England).

They all have a supreme leader who is responsible for "defending the faith" (statement of beliefs). The leader is also responsible for maintaining and monitoring the hierarchy in the corporate structure.

All the above-mentioned churches that operate in the United States must meet filling requirements with the I.R.S. to get a 501 (c)- (3) tax-exempt status. This is a highly prized designation among churches that donations in the form of tithes, gifts, or offerings from their members or the public may be written off on an individual's income tax returns. The rules to file for a 501 (c)- (3) tax exempt for a church-state that you must essentially be incorporated. By this, they mean that you must name Directors and Officers who have actual management authority over the affairs of the church. Officers are individuals who typically have authority with respect to the day-to-day operation of the corporation. Usually, they are referred to as President, Vice President, Secretary, and Treasurer of the corporation.

Directors (often referred to as "Trustees", "Elders", or "Board Members") usually have general management authority over the corporation. They are usually not involved in the day-to-day operations to the same degree as officers of the corporation. Instead, they have a right to vote on decisions, such as approving church budgets, changing church names, electing officers, amending the articles of incorporation or bylaws, borrowing money, designating compensation for the pastor, or dissolving the corporation etc.

In essence, the civil government is saying to be granted tax-exempt status, churches in the United States must be formed or incorporated like the Gentiles!

Matthew 20:35-27

25. But Jesus, Yeshua, called them unto Him, and said, you know that the princess of the Gentiles exercise dominion over them, and they that are great exercise authority upon them.

26. But it shall not be so among you: but whosoever shall be great among you, let him be your minister.

27. And whosoever will be chief among you, let him be your servant.

Although rules for incorporating churches are well intentioned, they lead the forming of Hierarchical Authority in churches. They also lead to a game of men playing politics, resulting in those who covet control (preeminence) constantly striving to climb the corporate ladder. A prime example of this is found in

III John 1:9-10

9. I wrote to the churches (the ecclesia), but Diotrephes who loved to have the preeminence among them receiveth us not.

10. Wherefore if I come, I will remember his deeds which he doeth, prating against us with malicious words, and not content therewith, neither does he himself receive the brethren, and forbids them that would, and casts them out of the church.

Unfortunately, the spirit of Diotrephes is alive and active in all too many churches today. Hierarchal Authority has caused grievous harm to the ecclesia and has disgraced the name of Yehovah, God the Father, and Yeshua, the Son of the Living God.

Let us examine the definition of the word "hierarchy"!

From the Dictionary of Oxford Languages.

Hierarchy: noun.

1. A system of organizations in which people or groups are ranked one above the other according to status or authority.

Example A: The initiative was with those lower down in the hierarchy.

Example B: The Roman Catholic hierarchy in the Vatican.

2. The upper echelons of a hierarchal system; those in authority.

Example: The magazine was read quite widely by some in the hierarchy.

Hierarchy Definition & meaning- Merrium Webster

Hierarchy: noun

1. A division of angels.

2. a. The ruling body of clergy organized into orders or ranks, each subordinate to the one above.

b. church government by hierarchy.

3. a body or person in authority.

4. The classification of a group of people according to ability or to economic, social, or professional standing.

So, you see, brethren, a hierarchal system that has granted leaders unchecked authoritarian control over the ecclesia in any organization is not conducive to a service-focused ministry as we see recorded in the primitive (original) New Testament Church.

In all truthfulness, this model in use today of hierarchical authoritarian control is nothing more than bowing to the Roman Catholic system.

I have witnessed firsthand and listened to horror stories about abuses inflicted on members in many different church denominations. The problem is that all too often, nobody is held accountable for the emotional harm that has been inflicted. The fear of being ostracized or even disfellowshipped causes most people to quietly pretend that there is no problem. Fear is a good motivator, but Yehovah God is not the author of fear.

II Timothy 1:7 For God is has not given us the spirit of fear, but of power, and of love, and of a sound mind.

The all-too-common denominator that causes leaders to be cavalier about their harsh judgements and authoritarian dictates is the assumption of status that leads to the feeling of ENTITLEMENT. Entitlement is a sneaky little devil that often creeps into churches that are based on the government-approved corporate model. The false ideas about hierarchy that stem from King James' translations of the Bible only lend to the problem. Another idea that enters the equation of thinking "we are special, and therefore entitled" is the teaching that we are The One True Church! We are the people of God, and you are not! Therefore, we pretty much keep to ourselves in our own little groups and have no dealings with those in the world. All too often, we look down our noses at "those people." This assumption of status with God and the feeling of entitlement will be the downfall of the Churches of God unless we repent.

Let's take it one step further. You have probably run into individuals, even in the one true church, who are even a little more "entitled" than most other members. Who are these special ones? Can you say Ambassador College Alumni?

Those who attended church college in Pasadena, California. Big Sandy, Texas, or Bricket Wood, England. My wife and I, while serving as door greeters at 3 different Feast of Tabernacles sites, have been snubbed by these extra special, super important, exclusively entitled individuals and their wives.

I can't help but think of the book Animal Farm. In this (revised) version of "Animal Farm" the animals of "Manor Farm" rebel against their oppressors and rename the farm "Animal Farm." They base their management on "Animalism" and one of the seven commandments, which is "All Animals Are Equal." The pigs who went to a special college consider themselves more intelligent, so they assume leadership roles. Visa-a-vis their entitlement, they begin to take more and more for themselves. One exceptionally charismatic pig chases off his opponents and forcefully takes control of the farm, naming himself. "Pigamus Maximus" Through an ingenious propaganda campaign, he posts signs everywhere declaring the New Supreme Commandment: "All Animals Are Equal, But Some Animals Are More Equal."

I hope you get a chuckle out of this little story. If you are a minister who came out of one of the previously mentioned systems, probably not so much.

I would highly recommend that everyone who is reading this book, The Malachi Matrix, listen to a sermon online by a true servant of God. The late Mr. Ronald L. Dart titled: "The Axe and The Fire." It deals with the whole issue of the assumption of status and entitlement and how dangerous this mindset is.

I would also recommend a paper titled: "Examining Church Governance" written by another humble servant of Yehovah, Mr. Will Burg. I will be quoting extensively from Will Burg's work because it is both scripturally and historically accurate.

Page #1. Paragraph 3-5.

Throughout human history there have always been individuals who wanted to rule over others, even in churches. Predicting that would happen in the New Testament assemblies of God's called-out people (Greek ecclesia), Paul wrote:

Acts 20:30 also from among yourselves men will rise up, speaking perverse things, to draw away disciples after themselves.

One way they did this is by elevating themselves above God's children and lording over them.

II Peter chapter 2 and III John 9&10. For one such example the brethren in Corinth were deceived by ambitious false teachers who led them into spiritual bondage.

II Corinthians 11:3-4, and 20.

3. But I fear, that by any means, as the serpent beguiled eve through his subtlety, so your mind should be corrupted from the simplicity that is in Christ Jesus.

4. For if he that cometh preaches another Jesus (Yeshua), whom we have not preached, or if you receive another spirit, which you have not received, or another gospel, which you have not accepted, you may well bear with him (you put up with him).

Verse 20. For your suffer, if any man brings you into bondage, if a man devours you, if a man takes of you, if any man exalts himself if a man smites you on the face.

Over the years, many have been spiritually and emotionally damaged by harsh, overloading, uncaring governance practices.

(stop)

So, this author asks you, brethren, have any of you been verbally abused by a minister? Have any of you been talked down to like you were an ignorant child? Have any of you been disciplined by having your speaking privileges revoked? How many have been made cast away by being kicked out of a church for asking questions or for quoting scriptures that conflict with the church's statements of beliefs? How many have been told you can believe what you want, just don't talk about it here? Freedom of speech, a cherished American right, is not allowed in some Churches of God. You don't dare discuss scriptures or question doctrines with other members without a minister being present, or you might be reported to the authorities and branded as a Heretic.

The King James translation of the Bible is not the infallible word of God that descended from heaven. In fact, it is loaded with intentional insertions, errors in translation, and words and phrases that were not in the original Greek or Hebrew.

Continuing with Mr. Will Berg's paper on "Examining Church Governance."

Page #1. Last paragraph.

Looking at more recent times, the rule and control of members by church leaders occurred, but in different more deceptive ways. For example, today, many Christians still accept and believe in the wording of the widely read 1611 King James Bible KJV., which contains hierarchical church organization and control terminology.

Page #2.

During the mid-20th century, when many Worldwide Church of God doctrines were being studied and established, the KJV. was the Bible most used for teaching and research.

Back in those days, few, if any, knew that the hierarchical words of the KJV are not in the New Testament Greek text. They were deliberately put in by men who wanted to validate and authorize hierarchical church practices.

In the Western world, the dominating religious influence throughout the Middle Ages was that of the Roman Catholic Church. This church created a hierarchical, top-down governance structure of Pope, Cardinal, Archbishop, Bishop, Priest, and Deacon.

The RCC. Use the vulgate Bible that was translated into Latin from Hebrew and Greek text by Jerome (345-419 ad). When he compiled it around 405 AD. It became the official Bible of the RCC. For the next 1000 years. While the Vulgate was superior to other translations of the time, the Dutch Catholic priest and theologian Erasmus (1466-1536) Uncovered over 600 textual errors in it, and the Oxford and Cambridge trained William Tyndale (1494-1536) found that the meaning of some Greek words were deliberately mistranslated to support the hierarchal teachings. When the King James version was being translated it was heavily influenced by the Church of England's Bishop Richard Bancroft, whose personal objective was to preserve the authority of the hierarchal governance structure of the Anglican Church governance teachings and practices of the Roman Catholic Church.

Bancroft had tight control over his hand-picked translators and was instrumental in keeping Vulgate governance wording in the KJV. Bible, as he shared King James' strong belief in the divine right and authority of Kings. King James even dissolved parliament for 10 years and ruled as an absolute monarch during that period.

Bancroft also convinced King James that the future of the monarchy related to and depended on the continual authority of the Anglican Church's hierarchical governance model.

Page #3.

Bancroft made sure that misleading translations of words supporting hierarchical governance doctrine and practices were retained in the King James version.

These irregularities were explained in much greater detail by Ulster E McGrath, Oxford Professor of Historical Theology, in his authoritative and very important book: "In the Beginning, 2001, Clarification of Words".

Correcting the deliberate KJV. Mistranslations are important, because words (especially if they relate to church governance) create wrong doctrines and practices that God never intended. Notice these examples:

Office - The KJV. Translators inserted the word "office" in five New Testament places to indicate someone who rules over God's people. Romans 11:13, 12:4, 1 Timothy 3:1,10,13. (see also Bishop and Deacon below). More accurate recent translations have omitted the word "office,

Bishoprick, and Bishop. These are KJV. Translations of the Greek word "episcopos" which means overseer a man who is appointed to watch over and serve the needs of the brethren in a Godly and orderly manner. (Acts 1:20 1 Timothy 3:1-2). He should be a righteous example who lovingly serves and shepherds God's people, rather than being an overlord (1 Peter 5:1-3) "presbeteros" which means "an older man" (Titus 1:5, 1 peter 5:1-3) it describes a responsibility and opportunity for wider service to the ecclesia, rather than a position of superiority or rulership. Presbuteros is like "episcopos" and is used interchangeably with it.

(Mounce's Complete Expository Dictionary, Words 2006 page 208)

An elder, overseer is to be a Godly example to the brethren, not a dictator or a tyrant. (I Peter 5:3) Catholics translate prebuteros as priest rather than elder. (The Great Ecclesiastical Conspiracy, Third Revised Edition page 18)

Deacon: From the Greek word "diakinos" which means servant. (Mounce page 159) like elder, it is not a position of hierarchical superiority or rulership.

Rule Over-Examining the KJV. Translation of Hebrews 13:7.

"Remember them we have the rule over you," the words "have the rule over you" are not in the Greek text, which reads "remember your leaders who spoke to you the word of God." (The Greek Interlinear New Testament) page 783.

"Rule Over" in the KJV is an intentional mistranslation of one Greek word, "hegomai" which means "to lead".

Adam Clark acknowledged this in his Commentary of this verse by translating it- Remember your guides, who have spoken unto you the doctrine of God. (Volume 6, page 786). Continuing in Hebrews 13, the KJV. Translators also changed verse 17 to read "obey them who have the rule over you," which intentionally endorsed the illegitimate hierarchal model of The Roman Catholic Church. This conflicts with Jesus, Yeshua's statement in Matthew 23:8- for one is your teacher the Christ, and you are all brethren, (not some above others).

The dictionary section in Mounces page 1,297 #5634 explains that the Greek word "obey" (hypakouo) means, "to give ear, to listen", which is a much less authoritative meaning than being commanded by church leaders that the word "obey" conveys.

(stop)

This is just a brief synopsis of the errors and intentional mistranslations in the widely accepted King James version of the Bible. Many churches, even to this day, cling to these errors to legitimize their hierarchical and authoritarian rule over the people of God.

All those who truly want to worship Yehovah, God the Father, need to learn a new way of lovingly serving the people of God. The old model, which is based on the Roman Catholic church, is not what God the Father intended.

The era of Big Corporate Churches and their authoritarian rule over God's people is coming to an end! If you are tired of being censored and not having a voice in the ecclesia, if you are no longer content to just pay and pray, I invite you to explore with me a new Bible-based model of equal partners in salvation. Learning to live and worship together with the divine spirit of love as our guide. Yehovah, God the Father, and Jesus, Yeshua, the Son of the Living God, are empowering INDIVIDUALS to Come Back Home! To be reconciled to Yehovah and Yeshua as they Ignite the Fire of REVIVAL among believers who are being called to worship in spirit and truth. The Father is planning a Homecoming! The Father is planning the wedding of His Son! If you have read this far, I believe the dynamic power of God's spirit is stirring in your soul.

Bear with me a little while longer as I lay out for you a vision of how "The Good Shepherd" gently leads the flock of God.

Matthew 11:28. Come unto Me, all you that labor and are heavy laden, and I will give you rest.

Announcing the Founding of Elijah Day of The Lord Ministries

Using this book, "The Malachi Matrix," as a catalyst, the time has come to begin a new ministry. The book of Malachi has awakened my spirit to begin a new journey of sharing this last-day message of turning the hearts of the fathers to the children and turning the hearts of the children to the fathers.

Malachi 4:4-6.

When I say "The Father," I am referring to the being who Jesus, Yeshua, called "My Father and your Father, My God and your God." John 20:17.

I truly believe that the spirit of The Living God has called me to serve Him in this ministry.

From this day and for the rest of my life, I will dedicate myself to preaching and proclaiming, Yehovah He is God!

Jesus, Yeshua, is the Son of the Living God. He pre-existed and was a servant and companion of Yehovah. He was an Elohim, a God being.

John 17:5 And now O Father, glorify thou Me with Thine own self with the glory I had with you before the world was.

Jesus, Yeshua, is "the Messenger of The Covenant" spoken of in the book of Malachi.

Malachi 3:1 Behold I will send My messenger, and he will prepare the way before Me, and the Lord whom you seek shall suddenly come to His temple, even the messenger of the covenant, whom you delight in: behold He shall come saith the LORD of hosts.

Jesus, Yeshua, divested Himself of the glory He shared with Yehovah, God the Father, and came down to this earth as a baby, born of Mary. He came to live as a flesh and blood human being so that He could experience and empathize with our frailty.

Jesus, Yeshua, manifested His Father's name, YEHOVAH.

John 17:61 I have manifested THY NAME unto the men which Thou gave me out of the world; Thine they were: and Thou gave them to Me, and they have kept Thy word.

John 17:26 And I have declared unto them THY NAME and will declare it: that the love wherewith Thou has loved Me may be in them and I in them.

Notice that experiencing the love of God the Father and being at one with Jesus, Yeshua is dependent on having the courage to declare the Father's name.

Jesus, Yeshua, boldly declared His Father's name and He will declare it for all eternity. In order to manifest and declare the Father's name, you have to KNOW the Father's

Matthew 6:9 After this manner therefore pray ye: Our Father which art in heaven, Hallowed be Thy name....

Psalm 83:18 That men may know that Thou whose name alone is JEHOVAH art the most high over all the earth.

I have recently heard two different Christian rock songs using the name "Yahweh." One song even referred to Jesus as "Yahweh." First, let me state categorically that "Yahweh" IS NOT GOD'S NAME!

The Hebrew scholar and Lecturer Nehemiah Gordon, in his book, "Shattering the Conspiracy of Silence," had this to say about the theory that God's name is Yahweh.

Page #69- (bottom of page)

The most popular theory that the name is Yahweh is based on a second-hand Samaritan tradition reported by 5th century Christion author named Theodorat of Cyrus, who didn't know Hebrew and was writing in Greek.

Do yourself a favor and read Nehemiah's book, "Shattering the Conspiracy of Silence." It will, God willing, convince you once and for all time that the proper name of the God of Israel is

YEHOVAH!

Finally, on this subject Jesus, Yeshua, is not Yahweh or Yehovah. God the Father and Jesus are Two Separate and Distinct God beings. Jesus, Yeshua, is currently seated at the right hand of Yehovah, God the Father, in heaven, in glory, and in honor, and in power.

Hebrews 12:2 Looking unto Jesus, Yeshua, the author and finisher of our faith, who for the joy set before Him, endured the cross, despising the shame, and is set down at the right hand of the throne of God.

Elijah Day of The LORD Ministries Mission Statement.

(Two Main Objectives)

#1. Elijah Day of The LORD Ministries is dedicated to preaching Malachi's message of RECONCILIATION to Yehovah, God the Father.

Malachi 4:4-6

4. Remember the law of Moses My servant, which I commanded unto him at Horeb for all Israel, with the statutes and judgments.

5. Behold, I will send Elijah the prophet before the coming of the great and dreadful day of the LORD, Yehovah.

6. And he shall turn the hearts of the fathers to the children and the hearts of the children to the fathers, lest I come and smite the earth with a curse.

#2. Reconciliation to Yehovah, God the Father, is only possible by accepting Jesus, Yeshua, as "The Lamb of God" to take away our sins.

Acts 2:36-39

36. Therefore let the House of Israel know assuredly that God hath made that same Jesus, Yeshua, whom you have crucified both Lord and Christ.

37. Now when they heard this they were pricked to the heart and said unto Peter and to the rest of the apostles, men and brethren, what shall we do?

38. Then Peter said unto them, repent, and be baptized every one of you in the name of Jesus, Yeshua, the Christ for the remission of sins, and you shall receive the gift of the Holy Spirit.

39. For the promises unto you and to your children and to all that are a far off, even as many as the LORD, Yehovah, our God shall call

Notice two things in verse 39.

A. "The promise is to you and to your children." Sounds very much like Malachi 4:6, turning the hearts of the fathers to the children and the hearts of the children to the fathers and turning all repentant hearts back to Yehovah, the one true God.

B. Not just Israel! "ALL That are far off, even as many as the LORD, Yehovah, shall call"! ALL Nations, kindreds, people, and tongues!

Revelation 7:9 After these things, I looked and behold a GREAT MULTITUDE which no one could number of all nations, tribes, people, and tongues, standing before the Lamb, clothed in white robes, with palm branches in their hands.

How is it possible there will be a Great Multitude in heaven? The only possibility is a mighty outpouring of the Holy Spirit from Yehovah, God the Father, in the last days, drawing millions who will choose to worship Yehovah, the One True God, and Jesus, Yeshua, the Lamb of God.

Joel 2:28-29 & 32.

28. And it shall come to pass afterwards, that I will pour out My spirit upon all flesh; and your sons and your daughters shall prophecy, your old men shall dream dreams, your young men shall see visions.

29. And also upon the servants and upon the handmaids in those days will I pour out My spirit.

Verse 32. And it shall come to pass, that whosoever shall call on the name of the LORD, Yehovah, shall be delivered: for in Mount Zion and in Jerusalem shall be deliverance as the LORD, Yehovah, hath said, and in the remnant whom the LORD, Yehovah, shall call.

WOW! A Last Day REVIVAL due to the outpouring of God the Father's spirit and the miraculous ministry of a thousand lights on a thousand hills. This multitude will respond to the calling of God the Father. They will repent and have their garments washed clean in the blood of the Lamb.

Joel 2:21 and 23-29

21. Fear not O land, be glad and rejoice for the LORD, Yehovah, will do great things.

23. Be glad then, you children of Zion and rejoice in the LORD, Yehovah, your God: for He hath given you the former rain moderately, and He will cause to come down the former rain and the latter rain in the first month.

24. And the floors will be full of wheat, (wheat harvest) and the vats shall overflow with wine and oil.

25. And I will restore to you the years that the locust has eaten, the canker worm, and the caterpillar, and the palmer worm, my great army which I send among you.

26. And you shall eat in plenty and be satisfied, and praise The Name of the LORD, YEHOVAH, your God, that has dealt wondrously with you, and My people shall never be ashamed.

27. And you shall know that I am in the midst of Israel, and that I am the LORD, Yehovah, your God and none else: and My people shall never be ashamed.

28. And it shall come to pass afterwards that I will pour out My spirit upon all flesh, and your sons and your daughters shall prophecy, your old men shall dream dreams, your young men shall see visions.

29. And also upon the servants and upon the handmaids in those days will I pour out My spirit.

Yehovah is no respecter of persons. The door of salvation be opened to all via The New Covenant.

Yehovah, God the Father, and His right-hand man, Yeshua, are moving heaven and earth and awakening men and women with the dynamic power of God's Holy Spirit. My heart is on fire, and my wife's heart is on fire with this vision we have of serving Yehovah and following Yeshua to His Father's House and to the Heart of the Father.

We are starting small, just my faithful wife Jackie and me. We are praying that Yehovah will send us fellow believers, a core group of 10 to 20 men and women who want to be part of this exciting ministry.

Elijah Day of The LORD Ministries!

We are using our tithe money to publish this book and praying that Yehovah and Yeshua will bless this labor of love. Most of all, we pray that this book will Give Glory to Yehovah's and Yeshua's Name. Proceeds from this book will be used to help pay for a website and recording equipment so that we can post sermons and Bible studies.

I am convinced there needs to be outreach in the form of a radio message. If Yehovah willing, we will be blessed with enough believers who want to form a fellowship. We will explore the possibility of filing for a 501 (c)- (3) tax exemption. We would only consider doing this if all the members of the fellowship (members consisting of men and women 18 or older who are baptized) vote to incorporate. We also will insist that all members of the fellowship be allowed to vote each year on who our servant leaders will be. If there is a tie vote on electing servant leaders or making other major decisions, we will allow each member of the fellowship to speak his or her mind, and then we will vote again. If we still have a tie after the second vote, we will draw lots.

No Baptized member in good standing will ever be disfellowshipped without a hearing before the entire ecclesia. If the infraction is proved to be egregious the members of the ecclesia will vote for disfellowship.

Following the example of former US Vice President Mike Pence, no man will meet privately with another woman other than his wife without at least two witnesses being present.

Scripture clearly pictures baptism as a public acknowledgment that an individual has repented of their sins and accepted Jesus, Yeshua's sacrifice as the Lamb of God to take away our sins, and that the individual now enters the New Covenant as a child of God the Father.

Luke 15:10. Likewise I, (Yeshua), say unto you, there is joy in the presence of the angels of God over one sinner that repents.

If the angels in heaven rejoice over one sinner who comes to repentance, you better believe we will rejoice here on earth with MUCH JOY!

We baptize in Jesus, Yeshua's name only.

Acts 2:38. And Peter said unto them repent and be baptized every one of you in the name of Yeshua Christ for the remission of sins, and you shall receive the gift of the Holy Spirit.

I am open to discussion as to the laying out of hands by elders (older men) of the fellowship at a baptism.

On a personal note, I am a motorcycle enthusiast. It is a personal dream to form a motorcycle club called: "The Sons of Elijah." This would enable men and women in the body of believers to bond and fellowship during road trips. I also hope to use this club to do good work in the form of helping widows and the fatherless. We could hold fund drives and deliver food, clothing, and other items to those in need.

The women in the body of believers need the fellowship and support of other women. To this end, we hope to form a women's group called: "The Abigail Sisterhood." They will hold Bible studies, be shaggers, or riding partners while delivering gifts and blessings along with "The Sons of Elijah."

The women will also hopefully be able to form a Sabbath School for our children and grandchildren.

I hope and pray you are inspired to join us on this journey. We will preach the eternal message of the Malachi Matrix, which is Reconciliation to Yehovah, God the Father. We will follow and emulate Yeshua, Jesus, The Good Shepherd. We will strive to become Able Messengers of the New Covenant! We will endeavor to live our lives in such a way that we become lights on a thousand hills, bringing Glory and Honor to Yehovah's name. We will uplift and support our brothers and sisters with the love of Jesus, Yeshua, in our hearts. We will pray for the day when Yeshua comes to rapture His bride and take us Home to meet the Father. We will be presented as the first fruits before the throne of Almighty God, Yehovah in heaven.

THIS WILL BE OUR HOMECOMING!

We will have the Father's name written on our forehead. Will be the bride of Christ. We will rejoice at the wedding supper of the Lamb. We will forever be the sons and daughters of the living God. We will bow in humble adoration and join the holy angels singing the song of Moses and the song of the Lamb.

Revelation 15:3-4

3. They sing the song of Moses the servant of God, and the song of the Lamb saying, great and marvelous are Thy works, Lord God

Almighty (Yehovah) just and true are Thy ways. Thou King of the Saints.

4. Who shall not fear Thee O LORD, Yehovah, and Glorify Thy Name? For Thou only art holy: for all nations shall come and worship before Thee: for Thy judgments are made manifest. Notice in verse 3, it mentions two songs. The song of Moses and the song of the Lamb. Who are they praising and singing about?

Lord God Almighty-Yehovah

Thou King of Saints- Yehovah

What will every nation on earth do?

Fear Thee O LORD-Yehovah!

Glorify Thy Name-Yehovah!

Worship Before God-Yehovah!

This is only the beginning of our journey. STAY TUNED!

If you are sincerely interested in being on the ground floor of forming a fellowship in the New Elijah Day of the LORD Ministries, please contact us at:

Elijah Day of The LORD Ministries C/O Richard Peterson

P.O. Box 611

Waterloo, Iowa 50701-0611

We would be happy to entertain any sincere request for interviews, guest appearances, book signings, or speaking opportunities.

We would also be happy to accept any contributions to help with our ministry.

We do not currently have a 501 (3)-(c) filing, so contributions are not tax deductible.

Please make all checks to:

Richard Peterson

P.O. Box 611

Waterloo, IA 50701-0611

The Message of Reconciliation!

The Ministry of The New Covenant!

The message of the MALACHI MATRIX is a message of Reconciliation to Yehovah, God the Father.

Come join us in Elijah Day of The LORD Ministries as we Bring Glory to Yehovah, God the Father, and Yeshua, the Son of The Living God!

Hebrews 12:22-24

22. But you are come unto MOUNT ZION, and unto THE CITY OF THE LIVING GOD, THE HEAVENLY JERUSALEM, and to an innumerable company of angels,

23. To the general assembly and church (ecclesia) of THE FIRSTBORN which are written in heaven, and to God (Yehovah) the judge of all, and to spirits of just men made perfect. 24. And to Jesus (Yeshua) THE MEDIATOR OF THE NEW COVENANT, and to the sprinkling that speaks better things than Abel.

AMEN AND HALLELUJAH!